Springer Tracts in Modern Physics 77

Ergebnisse der exakten Naturwissenschaften

Manuscripts for publication should be addressed to:

G. Höhler

Institut für Theoretische Kernphysik der Universität Karlsruhe

75 Karlsruhe 1, Postfach 6380

Proofs and all correspondence concerning papers in the process of publication should be addressed to:

E. A. Niekisch

Institut für Grenzflächenforschung und Vakuumphysik

der Kernforschungsanlage Jülich, 517 Jülich, Postfach 365

Surface Physics

Contributions by
P. Wißmann and K. Müller

With 58 Figures

Springer-Verlag
Berlin Heidelberg New York 1975

Privatdozent Dr. P. Wißmann

Institut für Physikalische und Theoretische Chemie,
Universität Erlangen-Nürnberg, 852 Erlangen, Egerlandstr. 3

Prof. Dr. Klaus Müller

Lehrstuhl für Festkörperphysik,
Universität Erlangen-Nürnberg, 852 Erlangen, Erwin-Rommel-Str. 1

ISBN 3-540-07501-1 Springer-Verlag Berlin Heidelberg New York
ISBN 0-387-07501-1 Springer-Verlag New York Heidelberg Berlin

Library of Congress Cataloging in Publication Data. Wißmann, Peter, 1936. Surface physics.
(Springer tracts in modern physics; 77). Bibliography: p. Includes index. 1. Metallic films – Electric
properties. I. Müller, Klaus, 1934. joint author. II. Title. III. Series. QC1. S797. vol. 77. [TN690]. 539'.
08s. [669'.95]. 75-30720.

Printing and bookbinding: Brühlsche Universitätsdruckerei, Gießen.

Contents

The Electrical Resistivity of Pure and Gas Covered Metal Films

By P. Wißmann. With 48 Figures

1. Introduction.. 1
2. Experimental.. 4
3. Structure of the Films... 13
4. The Resistivity of Pure Metal Films............................. 23
   4.1 Theoretical Treatment....................................... 23
       4.1.1 The Theory of Fuchs and Sondheimer.................... 24
       4.1.2 The Theory of Mayadas and Shatzkes.................... 25
       4.1.3 The Scattering Hypothesis............................. 26
       4.1.4 Comparison of the Theories............................ 29
   4.2 Experimental Results.. 33
5. The Temperature Dependence of Resistivity of Pure Metal Films. 44
   5.1 Theoretical Aspects... 44
   5.2 Experimental Results.. 46
   5.3 Limits of the Free Electron Gas Model....................... 52
6. Resistivity Change Due to Gas Adsorption........................ 57
   6.1 Theoretical Treatment....................................... 57
       6.1.1 Suhrman Model... 57
       6.1.2 Sachtler Model.. 59
       6.1.3 Fuchs-Sondheimer Model................................ 61
       6.1.4 The Scattering Hypothesis............................. 62
   6.2 Experimental Results.. 63
       6.2.1 Effect of Film Thickness.............................. 63
       6.2.2 Effect of Annealing Temperature....................... 67
       6.2.3 Effect of Measuring Temperature....................... 70
   6.3 The Maximum Value of Resistivity Increase................... 72
   6.4 The Discussion of the Over-all Curve $\Delta\rho(n)$............... 77
   6.5 Changes in Hall Constant and Thermoelectric Power......... 80
7. Resistivity and Heat of Adsorption.............................. 83
8. Concluding Remarks.. 86
References.. 89

How Much Can Auger Electrons Tell Us About Solid Surfaces?

By K. Müller. With 10 Figures

1. Introduction... 97
2. The Distribution N(E)... 98
3. Energy Levels of the Sample.................................... 99
4. Some Important Interactions.................................... 100
    4.1 Inelastic Loss Spectroscopy (ILS)......................... 100
    4.2 Disappearance Potential Spectroscopy (DAPS).............. 100
    4.3 Appearance Potential Spectroscopy (APS).................. 102
    4.4 Auger Electron Spectroscopy (AES)........................ 102
    4.5 Photoelectron Spectroscopy (XPS, UPS).................... 103
5. Instrumentation... 103
6. Auger Transitions... 105
7. Inspection of an Auger Spectrum............................... 106
8. The Qualitative Element Analysis.............................. 107
9. Steps Towards Quantitative Analysis........................... 109
10. Deconvolution.. 112
11. Line Shape and the Density of States......................... 113
12. Line Shape and Chemical Environment.......................... 115
13. Auger Electrons from Compound Solids......................... 117
14. The Concept of Cross Transitions............................. 118
15. A Potential Example: Cs-C.................................... 119
16. Metal Oxides... 121
17. Conclusion... 123
Abstract.. 123
Acknowledgement... 123
References.. 124

P. Wißmann

The Electrical Resistivity of Pure and Gas Covered Metal Films

1. Introduction

Resistivity measurements have received wide attention from experimental as well as theoretical side in order to gain insight into the electrical and structural properties of evaporated metal films lately. The relatively simple experimental arrangement and high accuracy are perhaps the main reasons for the popularity of this method. Motivating factors for the study of the thin film behaviour were many and varied such as: (a) Analyzing the thickness dependence of the film resistivity for getting information about the mean free path of conduction electrons /1/ (b) Developing integrated circuits, whose characteristics largely depend on the thin films used in their fabrication /2/ (c) Investigations of the "electronic factor" involved in heterogeneous catalysis by means of adsorption experiments performed on films produced under well defined UHV-conditions, where a clean film surface may be presumed to result /3/. Meanwhile, there is extensive experimental material at hand in the literature /4-8/. Nevertheless, it may be stated without any exaggeration, that an uniform interpretation of the results under review has not been possible till now.

Difficulties begin already with the explanation of the experimental results of the resistivity behaviour of the pure, uncovered films. They show specific resistivities which lie much higher than that of the bulk material. The thinner the film, the more marked is the effect. Size effect theory of FUCHS /9/ and SONDHEIMER /10/ has been used for the interpretation of this result in the literature nearly exclusively till now. According to this theory the conduction electrons suffer at the outer film surfaces at least a partially diffused scattering, which gives rise to an extra resistivity term especially at small thicknesses. The quantitative analysis of the experimental data shows, however, distinct deviations from the theoretical calculations /11/.

Table I: Fuchs Specularity p of Evaporated Au Films

reference	p
/12/	O - 1
/13/	O - 0,8
/14/	O
/15/	O
/16/	O - 0,9
/17/	O - 1

In other cases different authors have found extremly different numerical values for the characteristic Fuchs specularity of one and the
same material, as can be seen, for instance, for gold films in table I.
The values fluctuate between total specular (p=1) and completely diffuse (p=o) scattering. These fluctuations were correlated with the
influence of surface roughness on the scattering behaviour by most
of the authors /12-17/. Such an interpretation seems, however, to be
arbitrary. An effect of crystallite boundary scattering /18-20/ or
an influence of adsorption of foreign gas molecules, at least at the
boundary of film/substrate, can also play an important role. The results become specially vague when the resistivity measurements on the
polycrystalline gold films point at p=o, whereas optical reflection
experiments on the same films can only be explained on the precondition that p=1 /21,4/.

Many difficulties appear while attempting to explain adsorption
experiments too. Experimentally one always finds an resistivity increase with increasing coverage. This resistivity increase, which
becomes larger for decreasing thickness, was interpreted formerly by
a diminuation of the number of free electrons in the films. According
to the ideas developed first by SUHRMANN and co-workers /22/ the binding between adsorbent and adsorbate should be mainly polar, i.e.
there should be a direct transition of the conduction electrons from
the metal to the adsorbed molecules. This model, however, does not
exclude the possibility of electron displacements leading to a decrease in resistivity. Early works performed under relatively bad
vacuum conditions appear to confirm such a decrease indeed /23,24/.
Later on, however, it was shown, that at low coverages the resistivity increases in each case /25,26/. Besides this the thickness dependence given by the experiments shows a marked deviation from the

2

theoretical calculation based on the Suhrmann model /27/.

SACHTLER and co-workers /28/ proposed, therefore, to explain the resistivity increase as a demetallisation effect. The surface atoms of the adsorbent should form a surface complex with the adsorbed gas molecules and thus loose their metallic state to a certain extent. In this way the film thickness would be diminuated by approximately an atomic layer thus increasing the resistivity. With the help of this model a resistivity increase of about a few per cent could be explained, but the model fails completely in interpreting the 30 % increase observed in the case of CO-adsorption on about 100 $\overset{o}{A}$ thick copper films /29/. It remains ambiguous, too, as to why the CO relatively weakly bound on copper causes such a big resistivity increase in comparison to the case of nickel, where the CO is bound more strongly /3/.

Summarizing it may be stated, that models discussed in the literature so far do not give a complete picture for quantitative descriptions of the resistivity behaviour of clean and gas covered metal films. Aim of this article is, therefore, to demonstrate new possibilities for the discussion of the above mentioned resistivity effects taking into account the results published by the author in various papers in the recent past /30-39/. Mainly the properties of nickel and copper films shall be taken into consideration. These metals are particularly suited for the study of adsorption phenomena, because both have a face-centered cubic structure with nearly the same lattice constant, but differ in their electronic structure. Thus nickel shows, contrary to copper, a considerable number of unfilled holes in the d-band. Through comparison of the results obtained for these metals it should be possible, therefore, to answer, amongst other things, how far the assumption of a single band model is an admissible approximation in calculating thin film resistivity.

For these nickel and copper films it will be attempted to explain the experimental data under the assumption of a largely specular scattering of the conduction electrons at the undisturbed film surfaces and to trace back the observed thickness dependence of resistivity of the clean films to the scattering of the electrons at crystallite boundaries. The resistivity increase caused by gas adsorption can then be explained without any problem by a creation of new scattering centers at the film surfaces, similar to the scattering centers produced

3

by alloyed foreign atoms in the bulk. It will be checked, whether the influence of film thickness, annealing temperature, measuring temperature and gas coverage on the resistivity increase of the systems investigated here could be explained satisfactorily with the help of this hypothesis. Further on it will be attempted to take into account the temperature coefficient of resistivity, thermoelectric power and Hall effect measurements. In this way the limits of applicability of the suggested simple model can be made clear.

Results obtained on films other than nickel and copper are taken into account only in special cases because of the reasons given in section 2. The resistivity of very thin, discontinuous metal films /40,41/ cannot be discussed at all within the frame-work of the present article. Such films show strong structural and electronic characteristics, for the interpretation of which special models have to be developed /7/. With reference to the discussion of galvanomagnetic properties of the films like magnetoresistance effect /10,42,43/, anomalous skin effect /10,44/ and eddy current size effect /45/ we must refer to the corresponding literature too. The explanation of available experimental data, especially in the case of nickel and copper, is very complicated because simple models for the description of galvanomagnetic properties of the transition metals do not suffice as a rule /46-48/.

2. Experimental

A series of experiments on the resistivity behaviour of thin metal films has been performed by different authors on various systems, resulting in a large amount of experimental data on clean (see the review articles /1,4-7/) as well as on gas covered films (see, for instance, BLIZNAKOV and LAZAROV /49/, GEUS and ZWIETERING /8,50/, MURGULESCU and IONESCU /51/, COMSA /52/, PONEC and KNOR /53/, SACHTLER /28/, SUHRMANN /22/ and WEDLER /3/). On the first view, some of the existing data seem to show a rather poor consistency. The reason lies in the fact, that the films used by different authors often differ strongly in their structural properties due to varying deposition conditions.

In order to be able to compare the results of the resistivity measurements on uncovered films alone the following parameters must check: material, structure, temperature and purity of the substrate, rate and geometry of evaporation, residual gas pressure and composition, material, thickness and annealing treatment of the films, and measuring temperature. Or, at least it should be possible to estimate with sufficient accuracy the influence of the above mentioned parameters on resistivity behaviour. Obviously this is possible only in very few cases. High purity of the film surface and the used gases is necessary for adsorption experiments additionally, and an uniform distribution of the adsorbate on the surface must be assured. Therefore, the discussion of the experimental results in the present paper shall be limited to measurements which satisfy following conditions:

(a) films must have been evaporated and investigated under UHV-conditions;

(b) the structure of the films must be largely known;

(c) thickness, annealing temperature, measuring temperature and gas coverage should have been varied systematically, keeping all other parameters of preparation strictly constant;

(d) the gases to be adsorbed should be simple molecules such as CO, N_2 and H_2, in order to rule out the possibility of decay at the surface (like in the systems Ni/H_2O /54/ and Ni/CH_4 /55/, or of penetration into the metal (like in the system Ni/O_2 /56/);

(e) the coverage must be known to a sufficient accuracy;

(f) data on the absolute change of resistivity must be given, because data on the relative change only do not suffice for a quantitative comparison with theory.

In table II the papers are listed, which fulfil more or less the above mentioned conditions. Column 1 gives the system investigated, column 2 the author and column 3 the reference of the paper. Column 4 - 6 indicates, whether emphasis was given to thickness (d), annealing temperature (AT) or measuring temperature (MT) dependence, column

Table II: List of the Resistivity Measurements Discussed in Detail in this Paper

System	Author	reference	systematical variation of			additional investigation of			experimental device: fig. 1	F	f	literature for comparison
			d	AT	MT	TCR	TP	HE				
Ni/CO	WENZEL	/57/	+	-	-	-	+	-	b, (a)	1,2	1,4	/27,65,66/
Ni/CO	WISSMANN	/58/	-	+	+	-	-	-	a	0,83	1,8	/65/
Ni/H_2	REICHENBERGER	/59/	+	-	+	+	+	-	b, (a)	1,2	1	/65-69/
Ni/N_2	RICHTER	/60/	+	+	-	-	-	-	a	1*	1,2	
Ni	WÖLFING	/61/	+	+	+	+	-	-	a	0,83	1	/11,57a,71/
Cu/CO	KOCK	/62/	+	-	-	-	-	-	a	0,83	1,4	/29,70/
Cu/CO	WIEBAUER	/63/	+	+	-	+	-	+	c	1	1	/29,70/
Cu	RUDOLF	/64/	+	+	-	-	-	-	d	1	1	-

* In this case quadratical platinum foils were used, for which a calculation of the geometric factor was not possible. Therefore, F was set equal to one arbitrarily.

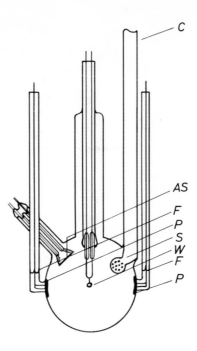

Fig.1a   Measurement cell for·the determination of resistivity of evaporated metal
          films. Details are given in text. The gas **shower** S **and** the additional sub-
          strate AS are not shown on their equatorial position for clarity reasons

7 - 9 informes, whether the temperature coefficient of resistivity
(TCR), thermoelectric power (TP) or Hall effect (HE) were investigated additionally. The literature references of those authors, who
have published comparable resistivity data on the same system, are
given in column 13. A comparison shows, that under the presumption of
similar preparation conditions of the adsorbent all results are in
surprisingly good agreement and have a remarkable reproducibility.

All the authors listed in table II investigated films evaporated
at a residual gas pressure less than $5 \cdot 10^{-10}$ Torr at a rate of about
10 Å/min. Glass held at 77K was used as substrate except in one case
/64/, where silicon single crystals were used. Thickness measurements
were always undertaken after completion of the resistivity investigation by dissolving the film and thereafter analysing quantitatively
with the help of different colorimetric methods /72/. In the paper /64/
thickness was determined additionally by continuous registration of
a vibrating-quartz-crystal monitor.

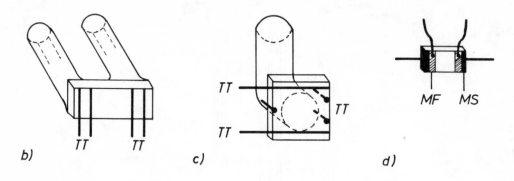

b)　　　　　　　　　c)　　　　　　　　d)

Fig.1b-d　Additional substrates according to WENZEL /57/, WIEBAUER /63/ and
RUDOLF /64/. TT = tungsten terminals, MF = molybdenum contact film,
MS = molybdenum sheet

The experimental arrangement is shown schematically in fig. 1. The
spherical glass bulb used in this or a similar form in all works is
represented by fig. 1a. This bulb provides the extra advantage of less
microroughness of substrate compared to the optically plane polished
glass substrates /73/.

The filament W, which consists of a spectrally pure nickel wire or
of small spectrally pure copper riders fixed on a spiral of a tungsten
wire, is situated at the center of the cell, providing an uniform film
on the inner wall of the bulb. Two ringformed platinum foils P are
fused on the inner walls facing each other, thus making possible resistivity measurements after connecting the foils with a Wheatstone
bridge by means of two respective feed-throughs F. The connection to
the vacuum pumpline is provided by C. The gas doses are given by break-
ing ampules, which were previously filled in a separate ultrahigh-
vacuum system with a known amount of the desired chemically pure gas.
If the cooling traps and the ionisation gauge are both separated from
the cell by manipulating from outside magnetically two glass valves of
ball-socket type, then it can be guaranteed, that the gas is adsorbed
nearly quantitatively on the film. Therefore, the coverage can be immediately represented by

$$n = \frac{\text{number of the adsorbed molecules}}{\text{cm}^2 \text{ of the geometrical surface}}$$

Sometimes a measurable equilibrium pressure develops over the film.
Then the n-values are provided with an adequate correction factor,
which becomes noticeable only at extremely high pressures (p > 10^{-4}

8

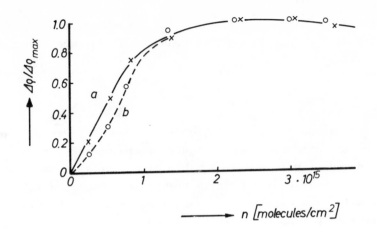

Fig.2 Resistivity increase due to CO-adsorption at 77 K. Gas shower in use (a)
and not in use (b) (WEDLER and FOUAD /27/)

Torr). A shower S serves the purpose of uniform distribution of the
gas on the film. That such a shower is necessary in cases where one
wants to study the resistivity change with respect to coverage of
immobile adsorbed gas molecules may be taken from fig. 2. In this
figure, the relative change $\Delta\rho/\Delta\rho_{max}$ of the resistivity of a nickel
film held at 77K is plotted against the carbon monoxide coverage /27/.
In the first stages of adsorption the gas is bound preferentially on
the equator region of the cell, if the shower is not used (o). This
results in a smaller increase in resistivity compared to the increase
with the shower in use (×). At 273K the CO-molecules would have sufficient mobility on the surface in order to compensate for a nonuni-
form distribution /27/.

In some cases (see table I column 10) an additional substrate AS
is installed in the cell. This may be a glass substrate with molten-
in tungsten contacts and polished optically plane surface (fig. 1b
and 1c) or a single-crystal silicon substrate with evaporated molybdenum contacts (fig. 1d). Details of the experimental arrangements
are described elsewere /57-64/, but in all cases the resistivity increase due to gas adsorption coincides within the experimental error,
whether measurement is performed on the spherical bulb or on a plane
substrate /57,59/.

In all the devices the main problem is the determination of the factors f and F, which are necessary for the calculation of specific resistivity of the film ρ from the measured resistance R according to the relation

$$\rho = \frac{d}{F \cdot f} \, R \tag{1}$$

Although the dimensionless /11/ geometry factor F is easy to be calculated for the different arrangements of the contacts shown in fig. 1a - d (numerical values in table II column 11), the determination of the correction factor f is troublesome, because it describes phenomena which are difficult both to measure or to calculate. This factor takes into account, that the fused platinum foils can loosen itself from the glass substrate indistinguishable for the eye (see fig. 1a), or that the transition between the glass surface and the tungsten terminal can be damaged by air bubbles, which might be opened due to polishing process (see fig. 1b and 1c), or that a part of the molybdenum contact film can lift itself during thermal treatment of the silicon substrate (see fig. 1d). In all these cases it is to be expected, that f takes up values greater than one.

Further it should be noticed, that the cylinder geometry model /11/ which is presumed for the calculation of the geometry factor F in the case of spherical cells, is applicable to the real situation only in the first approximation. For a more exact treatment the recesses left by the connections to vacuum system, gas shower, evaporation source etc. should be taken into account. Even if all these recesses are designed to be as near to the cell equator as possible, an estimation shows, that in an unfavourable case an error of about 30 % in the calculation of F must be reckoned with. This error should be assigned to a corresponding increase of the f-value too.

For an experimental determination of the correction factor f the product fρ calculated by (1) is plotted against the film thickness d for unannealed nickel films freshly evaporated at 77K (fig. 3). One can easily recognize the good reproducibility of the individual measuring points. A different f-value must generate a parallel displacement of the curves in such a double logarithmic plot. The figure shows, that this effect is really seen. If one chooses, as standard, the precision measurements of REICHENBERGER (□) /59/, obtained with the device of fig. 1b, then for all other series of measurements under consideration the correction factors listed in table I column 12 can

10

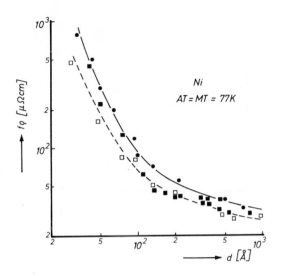

Fig.3 Resistivity ρ of nickel films freshly evaporated at 77 K as a function of film thickness d [WENZEL (●) /57/, REICHENBERGER (□) /59/ and WÖLFING (■) /61/]

be deduced. But it should be mentioned, that in unfavourable cases /71/ the magnitude of the f-values can take up to more than three.

The corresponding plot for copper films is shown in fig. 4, where this time the films were annealed at 293K. Again one recognizes, that the resistivity values decrease with increasing thickness. Here also a vertical displacement of the curves related to individual series of measurements is distinguishable. If one chooses the ρ-values of RUDOLF (▲) /64/ measured with the arrangement of fig. 1d as a standard, then one gets the f-values shown in table II column 12 for copper films.

Intentionally experimental conditions have been treated relatively in detail in this section. For, unfortunately, the comparison and the interpretation of the resistivity measurements of the published literature is complicated by the fact that the evaporation parameters of the films remain largely unknown in many cases. This led to a general opinion, that the resistivity change due to adsorption could be reproduced rather poorly and the results might be partially contradictory /4,7/. That this is not the case in practice can be deduced from the series of measuring points plotted in fig. 3, which were obtained by different authors. The relative deviation of the individual points

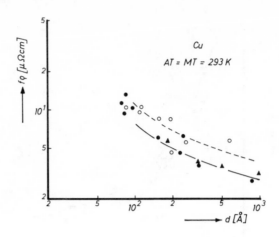

Fig.4 Thickness dependence of the resistivity of evaporated copper films.
[KOCK (o) /62/, WIEBAUER (●) /63/ and RUDOLF (▲) /64/]

from the plotted curve does not amount to more than 20 %. That deviations of this order of magnitude cannot be avoided in spite of keeping
all production parameterş mentioned at the beginning of this section
carefully constant, may have several reasons. Firstly, the thickness
determination carried out colorimetrically /72/ has a built-in accuracy of about 5 % only, and is related to a mean d-value without taking
into account the variations of film thickness over the entire surface
under investigation. Secondly, the evaporation process includes a few
parameters the influence of which has not found proper consideration
among authors doing this work. The rate of heating up of the substrate at the beginning of annealing process /30/ and a different pressure or composition of the residual gas due to a different purification
treatment given to the glass cell /74/, the filament /75,76/ or other
metal parts inside the cell /77/ may be counted amongst the above
mentioned parameters.

If the uncertainties related to the structure of the films could
be by-passed some how, then the reproducibility and the accuracy of
the results can be increased considerably. This is demonstrated in
fig. 5, where the relative resistivity change $\Delta R/R$ is plotted against
CO-coverage, using a 100 Å thick copper film as adsorbent. The film
was evaporated at 77K on a glass substrate and annealed for an hour
at 340K /62/. After it has cooled down to 77K again the film was exposed to gas coverage (×). One recognizes clearly, that the $\Delta R/R$-

12

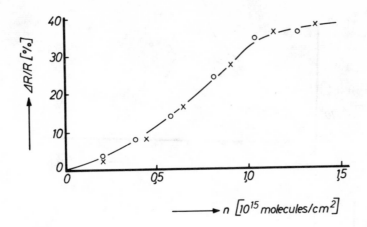

Fig.5 Resistivity increase of a 100 Å thick copper film at the first (×) and
second (o) cycle of CO adsorption (KOCK /62/)

values rise at first with increasing coverage and then achieve a saturation value before reaching the monolayer coverage. If, now, the
film is heated up to 273K once again, then the adsorbed CO is desor-
bed totally and can be pumped out from the gas phase. After recooling
to 77K the adsorption experiment can be repeated on the same film.
On performing this one gets values which are plotted additionally in
fig. 5 (o). Both the measurement series lie approximately on the
same curve providing an excellent check on the reproducibility.

3. Structure of the Films

From the discussion presented in the preceding section it is clear,
that for an interpretation of electrical properties of the films the
knowledge of their structure is of utmost importance. In this para-
graph, therefore, a short review of the results of structure investigations on nickel and copper films is to be given. If the metal to be
investigated is condensed under UHV-conditions on a glass substrate,

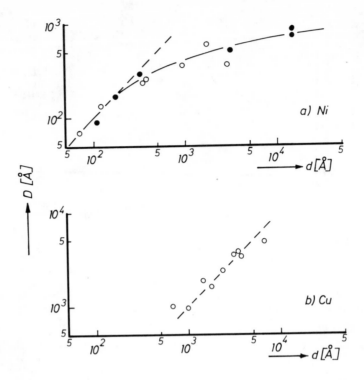

Fig.6   Mean crystallite size D as a function of film thickness d (AT = 293 K)
    a) Nickel films /31,79/   b) Copper films /78/

then polycrystalline films are produced which are built-up of many
individual crystallites. These crystallites show, in general, various
orientations and different shapes. Their averaged extension is cha-
racterized by the mean crystallite size, which can be determined
either by evaluation of the line - breadth of x-ray diffraction peaks
/31,78/, or by direct observation of the transmission electron micrographs /73,79/. The x-ray values characterize the extension of the
crystallites in a plane perpendicular to the film whereas the electron micrographs give their extension in the film plane.

Fig. 6a shows a plot of the mean crystallite size D for nickel
films evaporated at 77K and annealed at room temperature with respect
to film thickness d. A logarithmic scale has been chosen both for
abscissa and ordinate. One recognizes easily, that the D-values given
by x-ray measurements (empty circles /31/) and electron micrograph
evaluations (filled circles /79/) are in good agreement with the solid curve drawn additionally in fig. 6a. The crystallites possess,

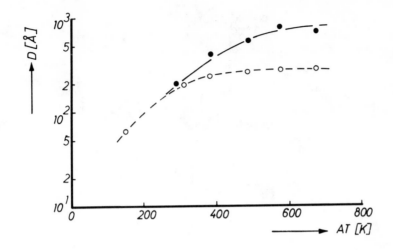

Fig.7 Mean crystallite size D of 250 Å thick nickel films as a function of
annealing temperature AT /31,79/

therefore, the same dimensions in both the directions. In other words,
they show - independent of the film thickness - an approximately cubic
shape. At very low thickness (below 150 Å), the crystallite size is
equal to film thickness. These films, therefore, are composed of one
crystallite layer only. For thicknesses above 150 Å the points lie
below the 45°-line, plotted additionally in fig. 6a (dashed line).
That means, the crystallite size is smaller than the film thickness.
Obviously the films now consist of more than one layer of crystallites, arranged one above the other.

The dependence of the mean crystallite size D on annealing temperature AT is shown in fig. 7 for a nickel film of about 250 Å thickness. Here a linear scale is chosen for the abscissa. The x-ray results are again presented by empty circles and the electron micrograph results by filled circles. One recognizes, that at low annealing
temperatures all D-values lie on one curve rising with increasing
annealing. For AT ≈ 450K the dimensions given by x-ray data reach a
limiting value which corresponds approximately to the film thickness,
while on the other hand the electron micrographs give much higher
D-values. Obviously the growth of the crystallites in the direction
perpendicular to the film plane is limited by the film thickness,
whereas the growth in the film plane can proceed uninhibited. In this
way the crystallites take up a laminar shape, as can also be seen
from the schematic representation of fig. 8.

15

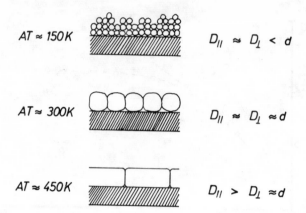

Fig.8 Schematic representation of crystallite growth in evaporated nickel films due to annealing

The measuring temperature has only little effect on the structure. Due to different thermal expansion coefficients of film and substrate a temperature change may cause either a development of a tension in the film or a removal of it /31/.

Results in agreement with the above observations are obtained for the structure of copper films too. However, the specific surface energy is so large in this case /80/, that above 340K a strong coagulation sets in causing the films to crack /62/. One observes according to fig. 6b, that already at room temperature the crystallite size equals the film thickness for all thicknesses under investigation. Obviously the relation

$$D \approx d \qquad (2)$$

illustrated in fig. 6b by a straight line, is valid in the present case. This relation has been proved by other authors in the case of aluminium films /20/ too, and seems to represent the normal behaviour.

Even under the best vacuum conditions it should be expected, that the films deposited at low temperatures contain a perceptible amount of built-in residual gas. A rough calculation shows, that at an evaporation rate of 10 Å/min and a residual gas pressure of $1 \cdot 10^{-10}$ Torr the film should be contaminated by a ratio of about 1 : 1000, if each gas molecule arriving at the surface remains stuck. The ex-

16

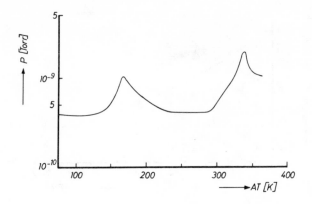

Fig.9 Desorption spectrum of a "clean" nickel film (WEDLER et al /65a/)

perimental proof of the built-in foreign molecules is obtained with
the help of desorption spectra typically shown in fig. 9 for a "clean"
nickel film /65a/. This approximately 400 $\overset{o}{A}$ thick nickel film was deposited at a residual gas·pressure of $2 \cdot 10^{-10}$ Torr at 77K and thereafter heated up slowly. One recognizes, that at certain temperatures
distinct pressure peaks appear which can be analysed with the help
of a mass spectrometer. It turns out, that these peaks can be traced
back to the desorption of N_2 nearly exclusively. N_2 is one of the
main components which the residual gas comprises of /76/. Obviously
the gas built-in by the evaporation process is displaced during crystal growth towards the crystallite boundaries and is partially desor-
bed. The other main component of the residual gas is CO. Since this
desorbs noticeably only at temperatures above 450K /65c/, it is to
be expected that even at room temperature the crystallite boundaries
still retain a certain amount of CO.

In the case of copper films, a complete desorption of CO at room
temperature should happen /3/. On the other hand, WILSON and SINHA
/81/ have demonstrated recently, that copper films deposited in an
argon atmosphere show higher resistivities than normal, which they
interpret by a noticeable inclusion of gas atoms in the films even
at room temperatur.

As can be deduced from fig. 8 qualitatively, the roughness of the
film surface should decrease with increasing annealing. This result
is confirmed, if roughness factor f_R is determined from adsorption

17

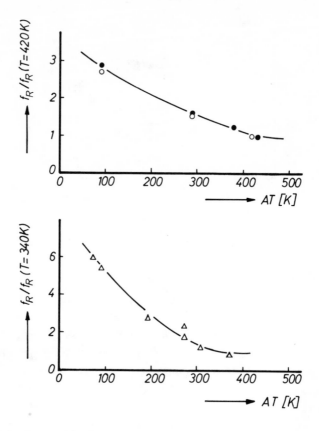

Fig.10 Effect of annealing on roughness factor f_R of nickel and copper films. Data obtained by RICHTER /60/ (o: Ni/N_2), WISSMANN /58/ (●: Ni/CO) and DAYAL /70/ (△: Cu/CO)

measurements[*]. f_R, which is defined by the relation

$$f_R = \frac{\text{true surface}}{\text{geometrical surface}}$$

is plotted against the annealing temperature for nickel /58,60/ and copper /70/ films in fig. 10. The value of f_R for highly annealed films is set equal to one arbitrarily, which seems reasonable considering the result of fig. 8. In fig. 10 it is shown explicitly that with decreasing annealing the roughness factor increases by a factor of three in the case of nickel and six in the case of copper.

---

[*] The coverage, at which a certain equilibrium pressure in the gas phase is established, serves as a measure for f_R in this case /27/.

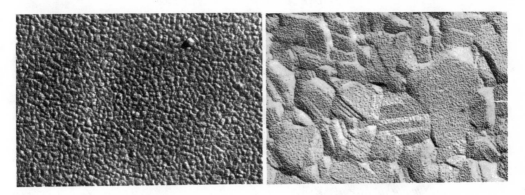

Fig.11 Replica of the surface of a 200 Å thick nickel film annealed at 293 K (a) and 673 K (b), respectively (GERDES /79/). Width of the photograph corresponds to 1,2 μ

A more detailed picture of the surface is obtained by using replica technique. Carbon-platinum films were deposited on the nickel film surface and than investigated in a transmission electron microscope after dissolving the nickel. Fig. 11 gives two photographs of such replica of differently annealed nickel films (d ≈ 200 Å) /79/. It is easily recognized from the photographs that no islands or holes develop in the film, and that the surface becomes more homogeneous with increasing annealing. For a nickel film of 250 Å thickness and annealed at 473K the evaluation of stereo photographs led to a roughness factor of f_R ≈ 1,1 /79/, which agrees well with the qualitative picture of fig. 8.

An increasing alignment of the individual crystallites towards (111) takes place simultaneously with crystal growth and smoothening of the surface. This is demonstrated by fig. 12a. Here a pole figure is presented, which was obtained by means of a x-ray texture analysis of the (111)-peak of a highly annealed nickel film /30/. One recognizes, that the intensity maximum falls at $\gamma = 0°$. The strongly broadend secondary maximum also stems from (111)-planes which are inclined to the film surfaces by $\gamma = 70,5°$. The highly annealed films, therefore, tend towards (111)-fiber texture, i.e. the individual crystallites are arranged with their (111)-lattice planes parallel to the substrate without any preferential azimuthal orientation.

If one follows the intensity along the axis A, then the texture profile given in fig. 13c is obtained /30/. One recognizes once more,

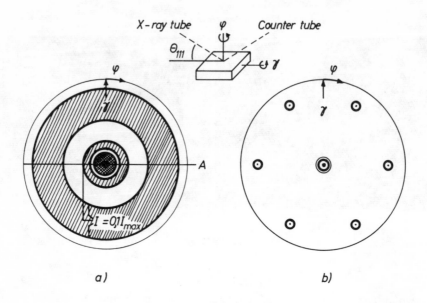

Fig.12 Pole figure of a polycrystalline (a) and single-crystal (b) nickel film
/30,58/

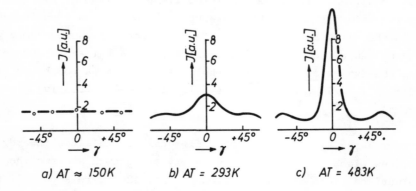

Fig.13 Texture profiles obtained along the axis A of fig. 12a for different
annealing temperatures (d ≈ 800 Å) /58/

that the extent of the preferential (111)-orientation in the highly
annealed film is specially marked. If the annealing is performed
only at room temperature, then a large number of crystallites of
other orientations is observed, and in the case of unannealed films
the distribution of crystallites of different orientation is completely random /58/.

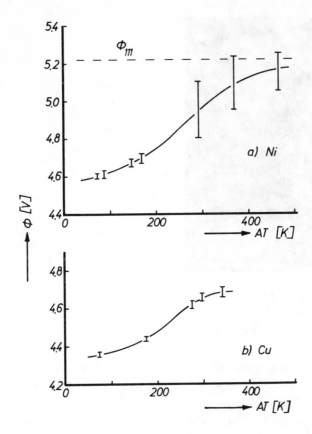

Fig.14 Change of work function φ with increasing annealing temperature AT for
        nickel (a) /35/ and copper (b) /62/ films

To make clear how far the texture profiles provide an indication
of the distribution of different indexed lattice planes <u>in the sur-</u>
<u>face</u>, the same films were investigated by low energy electron diffrac-
tion /75/. No measurable diffraction peaks could be observed in any
case. This result shows, that a remarkable disorder in the surface
structure is present even at the highest annealing temperatures,
for, in the other case the diffraction pattern should have shown a
characteristic (111)-ring maximum /82/.

Further important conclusions about the surface orientation can
be derived from the measurements of work function. In fig. 14a the
work function φ of clean nickel films is plotted against annealing
temperature /35/, and fig. 14b shows the corresponding dependence

Fig.15 LEED-pattern of a
(111)-oriented
single-crystal
copper film /58/

for copper films /62/. The ϕ-value **determined** for a (111)-oriented single-crystal nickel surface is represented by a dashed line additionally /83/. One recognizes that the unannealed films possess a relatively low work function which is in good agreement to the concept of high indexed surfaces or of steps and roughnesses in the surface /87/. The work function rises then with increasing annealing and approaches the value determined for a clean (111)-surface. Here, however, a certain fluctuation in the values is obvious, which points at a remaining disorder in the surface even in the highly annealed films.

The production of less disordered (111)-surfaces becomes possible, if one uses single-crystal substrates instead of amorphous glass. If one glows a (111)-oriented silicon crystal in ultrahigh-vacuum to approximately 1200 K just before starting the evaporation process, then nickel and copper films deposited at room temperature show to a large extent single-crystal structure /39/. Fig. 12b gives the pole figure obtained in the case of a 500 Å thick nickel film It is easy to discern the sharp secondary maxima, which points additionally at an appearance of azimuthal preferential orientation as compared to fig. 12a. On the same film diffraction patterns in a LEED-apparatus are found too /39/. Fig. 15 shows a typical photograph. The sharpness of the diffraction patterns indicates, that the surface structure has achieved a degree of order which is comparable to bulk single crystals.

If the silicon substrate is not glowed above 1100 K before evaporation, then the Auger electron spectrum of the silicon surface
shows a marked oxygen peak /84,75/. Obviously this oxygen hinders
the epitaxial growth of the films and no single-crystal structure
can develop. A similar result is obtained if evaporation procedure
was performed under bad vacuum conditions. In all these cases the
films show a more or less marked (111)-fibre texture similar to that
obtained for glass substrates.

4. The Resistivity of Pure Metal Films

4.1 Theoretical Treatment

As was mentioned in the introduction the calculation of the resistivity behaviour of pure metal films has been done mostly with
the help of size-effect theory of FUCHS and SONDHEIMER /9,10/. According to this theory a fraction (1-p) of the conduction electrons is
diffusedly scattered at the film surfaces, leading to an increase in
resistivity in comparison to the bulk. Unfortunately some simplifying
assumptions had to be made in the development of the theory, which
are difficult to realize or even cannot be approximated by experimental conditions. The assumptions in question are mainly the following:

(a) disorder in the films should be independent on film thickness;

(b) the films should have clean and plane parallel limiting sur-
    faces;

(c) the precondition of a single parabolic conduction band should
    be a permissible approximation;

(d) scattering process should take place isotropically and can be
    characterized by an effective mean free path of the electrons
    (which already implies the applicability of Matthiessen's rule

/85/), and by a single surface scattering parameter p.

That in case of nickel films at least the presumptions (a) and
(b) are not fulfilled at all can be seen from fig. 8 immediately.
Therefore, in this section we shall endeavour to discuss the theory
of MAYADAS and SHATZKES /86/ and a scattering hypothesis /33/ additionally, which allow a description closer to reality.

4.1.1 The Theory of Fuchs and Sondheimer

These authors solved the Boltzmann transport equation under the
assumptions (a) - (d). A simple boundary condition could be set up
following the plane parallelism of the film surfaces (or outer crystallite surfaces, see fig. 16a), which enables integration. The result is /1/

$$\rho_o/\rho = 1 - (3/2) \ x \ (1-p) \int\limits_{x}^{\infty} (1/s^3 - x^2/s^5) \ \frac{1-e^{-s}}{1-pe^{-s}} \ ds \qquad (3)$$

with $x = d/l_o$

where ρ and d are the resistivity and thickness of the film, ρ_o is
the resistivity of the bulk with same density of lattice defects as
the film, l_o is the corresponding mean free path of the electrons
and p is the fraction of electrons specularly reflected at the film
surfaces, i.e. scattered elastically without any change of momentum
component in the direction of the electrical field.

For sufficiently high thicknesses ($d \gg l_o$) eq. 3 simplifies to
/107/

$$\rho = \rho_o \ \left[1 + \frac{3}{8} \ (1-p) \ l_o/d \right] \qquad (4)$$

Lately many authors have tried to widen the region of applicability of the theory through suitable modifications. In this relation,
mainly, the case of different scattering parameters p and q due to
the transition film /vacuum and film/ substrate /88-90/, of nonspher-
ical Fermi surfaces /91-93/, and of an anisotropic mean free path
/94/ has been investigated. Special interest was shown for the study
of the dependence of scattering parameter p on incidence angle of
the electrons /95-100/ and on the roughness of film surface /97-102/.
Due to space limitation it is not possible to discuss these extensions

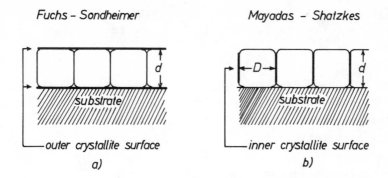

Fig.16 Outer and inner crystallite surface according to Fuchs-Sondheimer and
Mayadas-Shatzkes boundary conditions (schematically)

of the theory in detail here, literature mentioned and books /4/ or
review articles /5,6/ may be consulted for this.

4.1.2 The Theory of Mayadas and Shatzkes

Retaining the assumptions (b) - (d) these authors tried to describe
the thickness dependence of disorder by an one-dimensional grain boundary model /86/. This is achieved by modifying the boundary conditions
thus affecting the limits of integration of the Boltzmann equation in
comparison to Fuchs-Sondheimer theory. Under the assumption that the
individual crystallite boundaries are separated by a mean distance D
(see fig. 16b), the film resistivity is calculated to be

$$\rho_o/\rho = 3 \left[(1/3 - \alpha/2 + \alpha^2 - \alpha^3 \ln (1 + 1/\alpha)) \right] \tag{5}$$

with $\alpha = \dfrac{R}{1-R} \cdot l_o/D$

which for $\alpha \ll 1$, that is sufficiently large crystallites, changes
into

$$\rho = \rho_o \left[1 + \frac{3}{2} \; \frac{R}{1-R} \; l_o/D \right] \tag{6}$$

In these equations ρ_o and l_o have the same meaning as in (3). R corresponds to Fuchs specularity p and means, here, the fraction of elec-

trons reflected specularly at the grain boundary (or inner crystallite surfaces, see fig. 16b), through which the fraction 1-R travels
undisturbed.

Mayadas and Shatzkes also calculated resistivity behaviour for the
case of both grain boundary and films surface scattering. These calculations lead, however, to complicated integrals, which could be evaluated only numerically /86/.

4.1.3 The Scattering Hypothesis

Starting point is the relation /33/

$$\Delta \rho^* = \frac{mv}{Ne^2} \; A^* \; c^* \quad , \tag{7}$$

which has proved to be quite useful for the discussion of the effect
of lattice disorder /103/ and built-in foreign atoms /104/ on the
resistivity of bulk metals, even in the case of transition metals.
In this equation $\Delta \rho^*$ stands for the resistivity increase caused by
lattice disorder, mv is electron momentum at the Fermi surface, N is
the number of free electrons per cm^3, A^* is the mean cross section
and c^* is the concentration of the scattering centers per cm^3. The
star has been used to indicate that here primarily lattice defects
inside the films and not at the surfaces are considered.

If one identifies the crystallite boundaries in the films as disorder region which is equivalent to a row of "smeared-in" lattice
gaps /105/, or which even possesses built-in gas molecules /65a/,
then (7) should be applicable in this case immediately. But it should
be borne in mind, that using (7) the presumptions (c) and (d) must
hold /104/, while the presumptions (a) and (b) are insignificant.

For the sake of simplicity one assumes analogous to Matthiessen's
rule, that the total resistivity ρ comprises additively of two terms.
Thus

$$\rho = \rho_o + \Delta \rho^* \tag{8}$$

holds, where ρ_o is the resistivity of the undisturbed film and $\Delta \rho^*$
the resistivity due to lattice disorder. Considering at first only

the influence of electron scattering at grain boundaries or <u>inner</u> crystallite surfaces (see fig. 16), one has to set for c^* in (7) the concentration of the scattering centers in the grain boundary, and for A^* the mean scattering cross section. Putting the number of scattering centers per cm^2 of the inner crystallite surface as z^* it follows

$$c^* = z^*/D \tag{9}$$

Setting this in (7) leads to

$$\Delta\rho^* = (mv/Ne^2) \; z^* \; A^*/D \tag{10}$$

or, taking into consideration the Sommerfeld relation

$$\rho_o = (mv/Ne^2) \cdot 1/1_o \tag{11}$$

and the additivity condition (8), we have

$$\rho = \rho_o \left[1 + z^* \; A^* \; 1_o/D\right] \tag{12}$$

In this derivation it has been implicitly presumed, that the lattice inside the individual crystallites has no defects. It can, however, be easily shown, that if the lattice disorder inside the crystallites is independent of crystallite size, then (12) remains unchanged by choosing a suitable value of ρ_o. Furthermore (12) can be made to hold again by proper choice of correction factor for the product $z^* \; A^*$, if the lattice disorder changes inversely proportional to the crystallite size D.

Taking a step further we shall try to extend our scattering hypothesis also to electron scattering at the outer crystallite surfaces, i.e. at the film surfaces (refer to fig. 16). In agreement with the experimental results mentioned already in the introduction /12,13,17,21/ we shall presume, that the conduction electrons are reflected more or less specularly at the <u>ideal</u>, undisturbed metal surfaces. In this case, since the electrons do not suffer a noticeable change of momentum in the direction of the field ($M_{\parallel} = M'_{\parallel}$, refer to fig. 17a), the conditions at the interface film/vacuum correspond to the free motion of the electrons in a film of greater thickness.

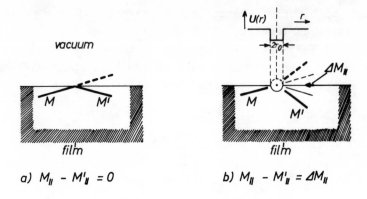

Fig.17 Scattering of conduction electrons at surface defects (schematically)

If the surface contains dislocations, stacking faults or foreign atoms, then a charging up at these defects is to be expected, similar to that found at a defect in the bulk of the metals. This charge must be neutralized by the conduction electrons in its vicinity. Such a defect can be described by means of a semispherical shaped, screened Coulomb potential, the radial distribution of which in a first approximation can be represented by a rectangular profile of the width r_o /106/. This potential distribution is indicated in fig. 17b, but for clarity reasons r_o is drawn there on a magnified scale. In the case of metals the actual radius of the screened potential should be about 10^{-8} cm /106/. From fig. 17b one recognizes immediately, that due to the semispherical shape of the potential well this model takes care of the disturbed three-dimensional symmetry /109/ to a certain extent. On the other hand, this model yields a momentum change $\Delta M_{\parallel}$ parallel to the direction of field (only this component is important for the calculation of resistivity), which complies exactly with the momentum change resulting from electron scattering at a spherically symmetric potential /110/ inside the bulk (refer to fig. 17b). In other words, it should be possible to describe the influence of defects at the outer film surfaces by means of an equation corresponding to (7), presuming always a specular reflection of the conduction electrons at the undisturbed film surface.

If c stands for the concentration of the scattering centers at the film surface and A for the related mean scattering cross section,

28

then analogous to (7) and (9)

$$\Delta \rho = (mv/Ne^2) \; A \; c \tag{13}$$

and

$$c = Z/d \tag{14}$$

should hold, where Z is the number of scattering centers per cm^2 of the film surface, and d is film thickness. Setting (11), (13) and (14) in (8) leads to a formula* quite similar to (12)

$$\rho = \rho_o \left[1 + Z \; A \; l_o/d\right] \tag{15}$$

If the electrons are scattered at the inner as well as at the outer crystallite surfaces, than according to

$$\rho = \rho_o + \Delta \rho_{inner} + \Delta \rho_{outer} \tag{16}$$

the following must hold

$$\rho = \rho_o \left[1 + z^* \; A^* \; l_o/D + ZA \; l_o/d\right] \tag{17}$$

4.1.4 Comparison of the Theories

Comparing the formulae given in the preceding section for the thickness dependence of film resistivity, it is surprising to note that (15) derived in a very simple manner corresponds to (4) calculated for $d/l_o \gg 1$ by the theory of Fuchs-Sondheimer. If one connects the characteristic constants according to the relation

$$Z \; A = 0.375 \; (1-p) \tag{18}$$

then (15) goes over to (4). Also a comparison of (15) with the exact Sondheimer relation (3) shows practically no deviation in the thick-

---

* It should be mentioned, that this equation differs only slightly from the well - known Nordheim formula /108/

$$\rho/\rho_o = 1 + l_o/d$$

The difference consists only in introducing an additional scattering parameter ZA in the case of (15).

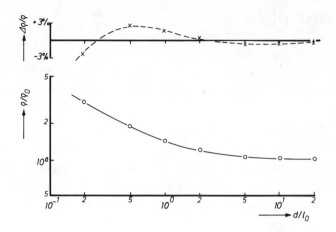

Fig.18 ρ/ρ_o (refer to (15)) and $\Delta\rho/\rho$ (refer to (20)) as a function of film thickness d

ness dependence when the connecting equation

$$Z\ A = 0.435\ (1-p) \tag{19}$$

is chosen. That is demonstrated in fig. 18. Here the correlation between resistivity ρ and thickness d is plotted in a double logarithmic way according to (3) choosing for the sake of simplicity p = o. If one were to plot in the same figure the curve given by (15) and (19), then there would be no recognizable deviation. In order to illustrate this point fig. 18 contains additionally, on the same abscissa, the relative difference between both function values

$$\frac{\Delta\rho}{\rho} = \frac{\rho(3)\ -\ \rho(15)}{\rho(3)} \tag{20}$$

Obviously, the deviations in the region

$$d/l_o > 0.2 \tag{21}$$

are less than 3 %. Only this region is of interest, however, to a comparison with experimental data. For, it is known from the investigations of TCR /11/ or work function /35/, that nickel films of thicknesses less than a critical thickness of about 40 Å crack or show, at least, a quite abnormal electrical behaviour, which can be attributed to an island structure /40/. At the same time a mean free path of the order of 133 Å is to be expected at 273 K /11/, so that con-

ductivity measurements on the continuous films are meaningful only
for thicknesses in the range $d/l_o > 40/133$. Thus, the condition (21)
is satisfied thoroughly. A similar discussion holds for copper films
which have a higher l_o-value (about 387 $\overset{o}{A}$ at 293 K /4/) but crack on
the other hand at a higher critical thickness (75 $\overset{o}{A}$ /62,70/).

Taking into consideration, that the thicknesses cannot be determined in general to an accuracy better than 3 % and that the thickness
enters according to (1) quadratically in the final expression

$$R = (\rho_o fF/d) \quad (1 + ZA \, l_o/d) \tag{22}$$

it becomes obvious, that the deviations presented in fig. 18 lie within the experimental error.

For the description of electron scattering at the grain boundaries
(12) should be used. This equation goes over into (6) calculated by
Mayadas and Shatzkes for sufficiently large crystallites, so far as
one connects the characteristic constants according to the relation

$$z^* A^* = 1,50 \, R/(1-R) \tag{23}$$

The graphical representation of (12) is given in fig. 19. In this
case too, the difference from the exact Mayadas function (5) is
negligibly small, if one chooses a suitable connection of the constants. For
$$z^* A^* = 1,35 \, R/(1-R) \tag{24}$$

the relative difference of both function values

$$\frac{\Delta\rho}{\rho} = \frac{\rho(5) - \rho(12)}{\rho(5)} \tag{25}$$

is plotted in fig. 19 additionally, it carries a value less than
1,5 % for

$$\frac{1-R}{R} \frac{D}{l_o} > 0.002 \tag{26}$$

This is an accuracy, which cannot be achieved in an experimental determination of mean crystallite size in any way /31,73/. The non-
fulfillment of the condition (26) means, according to fig. 19, ρ/ρ_o-

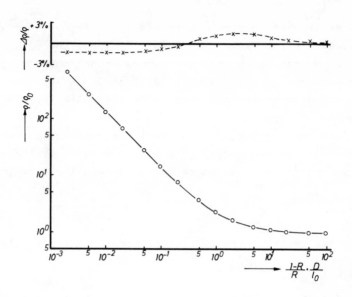

Fig.19 ρ/ρ_0 (refer to (12)) and $\Delta\rho/\rho$ (refer to (25)) as a function of mean
crystallite size D

values above 600. Such high resistivities can be obtained only in
the case of films with island structure, for which the models discussed here are not applicable at all /7/.

In the case in which the crystallites posses an approximately
cubic shape so that (2) is valid, (17) changes into

$$\rho = \rho_0 \left[1 + (ZA + Z^* A^*) \, l_0/d \right] \qquad (27)$$

This relation corresponds to (15) largely and it is immediately
obvious, that a separation of the both involved scattering processes
cannot be performed by the measurement of thickness dependence alone.
On the other hand, a quantitative calculation of the quantities Z,
Z^*, A and A^* is impossible for the time being, because no sufficient
information is available for the number and the properties of the
individual scattering centers. Therefore, other data such as the resistivity change caused by gas adsorption (see section 6.3) should
rather be taken into account in order to get more information and to
come to more definite statements.

(27) has another very important consequence. By comparison with
experimental data it is possible, no doubt, to check its applicabili-

32

ty as far as the thickness dependence of the resistivity is concerned. But on the other hand no conclusions can be drawn about the magnitude of l_o. With that an essential goal of earlier works, that is, to experimentally determine the mean free path of the conduction electrons by resistivity measurements on thin films /1/, has been put into question once again.

Summarizing, it may be stated that the thickness dependence of resistivity calculated with the help of the scattering hypothesis does not differ much from that calculated with the help of Fuchs-Sondheimer and Mayadas-Shatzkes theories. In the following, therefore, we shall utilize for interpretations only the relations (12) and (15), which remain valid in the presence of rough crystallite surfaces also.

4.2 Experimental Results

Before attempting any comparison of the experimental results with theoretically calculated values we must point out, that just in the case of nickel and copper films the correctness of certain assumptions underlying all the theories is disputable /109/. This concerns specially the assumption of a free electron gas, in which the scattering process is described by only one effective mean free path l_o and a single effective cross section ZA. For it is known from investigations on bulk metals, that such a simple model can describe only the resistivity behaviour - within certain limits - with sufficient accuracy /103,111,112/. On the other hand, for the interpretation of other electrical properties like temperature coefficient of resistivity, thermoelectric power or Hall constant /112-116/ more complicated factors like band overlapping, non-spherical Fermi surfaces, anisotropic scattering /111/ and electron spin coupling /117/ should be taken into account. Due to this, it is to be expected, therefore, that in order to discuss the thin film behaviour only the resistivity measurements can be interpreted with the help of the theories described in section 4.1, whereas for the interpretation of other electrical properties the true band structure must be taken into consideration. So we shall limit our discussion at first only to the resistivity of the nickel and copper films. In the case of nickel one should imagine while using a single-parabolic-band model, that here essentially the s-electrons are responsible for conductivity by suffering extra scattering at the holes of the d-band /111/. The influence of a form of

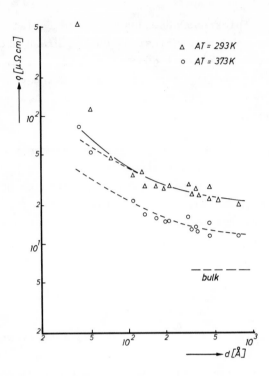

Fig.20 Thickness dependence of resistivity of nickel films (MT = 273 K, WÖLFING
/61/) —— computed from (12) --- computed from (15)

the Fermi surfaces more close to reality may be taken into account
by introducing an effective electron density n_{eff} /112/, but it can
be easily shown, that this quantity n_{eff} cancels itself in the derivation of (12) and (15).

In a later chapter (5) we shall point out a few paradoxes, which
result on the application of the very rough free electron gas model
for the interpretation of more complicated electrical properties of
nickel and copper films. Limitations of this simple model will be,
then, more distinct.

Fig. 20 shows the thickness dependence of the resistivity of evaporated nickel films measured by WÖLFING /61,36/. The films were
deposited on a glass substrate held at 77 K. After finishing the evaporation process the films were annealed at 293 K (Δ) and 373 K (o),
respectively, for an hour, and finally resistance measurements were

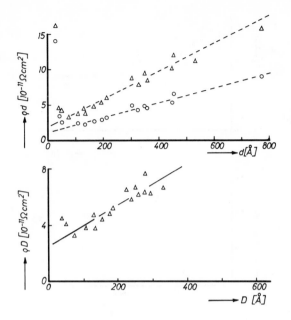

Fig.21 Plot of the points of fig. 20 in order to prove (28)

performed at $MT^* = 273$ K. Both axis are on a logarithmic scale to provide a direct comparison with the theoretical curves shown in fig. 18 and 19. Such curves are plotted additionally in fig. 20 using for calculation (15) (dashed line) and (12) (solid line). Obviously the theoretical curves and the experimental values coincide only for higher film thicknesses, whereas for $d < 70$ Å the measured values lie too high. One recognizes, however, that the curve calculated from (12) shows a larger radius of curvature and describes the experimental values a little bit more closely.

The constants ρ_o, ZA and $Z^* A^*$ needed for the calculation of the theoretical curves were taken in a preceding step from a plot ρd against d and ρD against D, respectively. Fig. 21 shows these plots for the same measured values as in fig. 20. According to the relations

$$\rho d = \rho_o d + ZA\rho_o l_o$$

---

* MT stands for measuring temperature

35

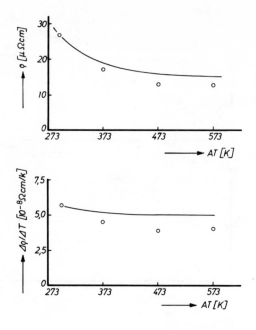

Fig.22 ρ and $\Delta\rho/\Delta T$ of a 200 Å thick nickel film as a function of annealing temperature AT (MT = 273 K) /58/. Solid curves computed from (12)

and   (28)

$$\rho D = \rho_o D + z^* A^* \rho_o l_o$$

which can be easily derived by transforming (15) and (12), the experimental values must lie on straight lines. Fig. 21 veryfies this, in fact, for large thickness ranges (in the case of fig. 21b the thickness dependence of the mean crystallite size D given in fig. 6a was taken into account). The constants determined from the axis intercept and the slope of the straight lines shown in fig. 21 are presented in table III.

(12) can also be used for the description of the dependence of film resistivity on annealing treatment. The resistivity must decrease as the mean crystallite size increases with increasing temperature. This decrease is observed indeed (fig. 22a), but the values measured for nickel films lie distinctly lower than the curve computed with the help of the experimentally determined relationship

Table III: Characteristic Constants of Thickness Dependence of
the Resistivity of Evaporated Nickel and Copper Films

| Metal | AT^* [K] | Reference | ρ_o [$\mu\Omega$cm] | evaluated with the help of equation | | |
				(12) $z^* A^* l_o$ [Å]	(15) $z A l_o$ [Å]	(18) l_o [Å] for p=o
Ni	293	/36/	13	220	–	–
	293	/36/	19	–	125	330
	373	/36/	9,5	125	125	330
	bulk	/11,118/	6,14	–	–	133
Cu	293	/64/	2,9	–	130	350
	293**	/64/	1,9	–	<27	<75
	bulk	/4/	1,67	–	–	387

*   AT stands for annealing temperature

** Single crystal film (refer to fig. 23b)

D=D (AT)* (taken from fig. 7) and the numerical values of table III.
Obviously, here we have to take into consideration, that with increasing annealing the disorder _inside_ the crystallites may decrease.
The reasons for this could be a healing up of stacking faults /80/,
the formation of a preferred orientation /30/ or a reduction of ten-
sile stress /119/. It is also imaginable, that the number of scattering centers decreases due to removal of gas atoms sitting in-be-
tween the grain boundaries /65a/. That different mechanisms are involved, can be deduced from the numerical values of table III. For
the highly annealed films (AT=373 K) one observes a distinct lowering
of the ρ_o-value and at the same time a decrease of the $Z^* A^*$-value
compared to the results obtained for the less annealed films (AT= 293 K).

The thickness dependence of the resistivity of copper films is
given in fig. 23. Fig. 23a first of all shows the values obtained by
WIEBAUER /63/ for films evaporated on glass substrates (DT**=298 K,
AT=330 K, MT=273 K). Additionally, a curve computed according to (15)
is plotted, for which the constants were determined as explained
above. Again it is noticed, that for thicknesses less than a critical
threshold, which is about 200 $\overset{o}{A}$ in the case of copper films, the experimental values lie higher than expected according to the theory.
An evaluation with the help of (12) leads to the same discrepancy,
because in the case of copper films approximately D ≈ d can be taken
from fig. 6b.

Similar results are obtained for films deposited on singel-crys-
tal silicon - 111 - substrates. Fig. 23b illustrates the electrical
behaviour of copper films investigated by RUDOLF /64/ (DT = AT = MT =
298 K). Obviously, now two different cases are to be distinguished
with respect to the thickness dependence:

(a) If films are deposited on unglowed substrates, then a marked
    thickness dependence can be detected even at higher thicknesses.
    This is indicated by the curve through the filled circles in
    fig. 23b. This behaviour complies more or less with that of films
    deposited on glass substrates.

(b) If the silicon substrate is glowed to a temperature of about 1200 K

---

* AT stands for annealing temperature
** DT stands for deposition temperature

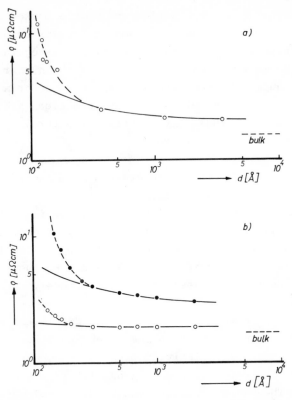

Fig.23 Resistivity ρ of copper films as a function of film thickness d
a) glass substrates, DT = 298 K, AT = 330 K, MT = 273 K (WIEBAUER /63/)

b) silicon-(111) substrates, DT = AT = MT = 293 K (RUDOLF /64/)

Solid curves calculated with the help of (15)

in vacuum immediately before starting the evaporation process, then the thickness dependence above the critical threshold of d = 200 Å disappears completely (empty circles in fig. 23b). The resistivity is only a little more than the bulk value, then.

Interpretation of this behaviour of copper films may be undertaken as follows:

On the basis of the results of Auger electron analysis mentioned in section 3 it is assured, that unheated silicon crystals have a noticeable amount of built-in oxygen in the surface /75,84/. From x-ray and low energy electron diffraction analysis it is known, that in this case a polycrystalline structure develops (refer to section 3)

with high density of grain boundaries and with a marked surface roughness (perhaps with occluded oxygen atoms in the surroundings of the
interface film/substrate). All these structural irregularities can
be interpreted by means of an augmentation of the number of scattering
centers thus producing higher values of ZA. Only after a sufficient
heat treatment of the substrate resulting in a clean silicon surface
evaporation leads to a single crystal film structure, by which the
ZA-value is lowered practically to zero.

The characteristic constants determined from the experimental
values of fig. 23b are additionally put into table III. It is striking
to note, that as in the case of nickel films, the formation of the
better ordered structure with a smaller density of grain boundaries
is accompanied with a decrease of resistivity ρ_0 inside the crystallites. The single crystal copper film shows a ρ_0-value only about 15 %
higher than the bulk·value (see table III column 4).

Unfortunately here we cannot consider in detail the extensive experimental evidence accumulated in the recent past for the tickness
dependence of electrical resistivity of nickel, copper and other
metal films. We must refer to the original literature and some review
articles /4-6/. The various results can be discussed more or less on
the above lines, but a few details seem to be worth remarks:

(a) By evaporating and annealing at the same temperature of 343 K
    largely undisturbed nickel films could be produced even by using
    glass substrates. For such films EYRICH /120/ found a ρ_0-value
    only about 35 % higher than that for the bulk. Obviously, here
    the deposition temperature was sufficiently high to hinder par-
    tially an inclusion of carbon monoxide always present in the re-
    sidual gas, and to enable the nickel atoms to occupy places of a
    better degree of order thus providing advanced crystallite growth.

    On the other hand, copper films deposited at room temperature at
    a residual gas pressure above 10^{-8} Torr /63/ or at an argon par-
    tial pressure of about 10^{-6} Torr /81/ showed slightly higher
    ρ_0-values than the ultra-pure films.Here the results point at a
    formation of lattice defects caused by the foreign gas molecules
    included during evaporation. This is a surprising effect, because
    Ar as well as the main components of the residual gas (N_2 and CO
    /76/) cannot be adsorbed normally on copper surfaces at room tem-

perature /3/.

(b) A resistivity nearly equal to the bulk value, which remains constant down to thicknesses of about 250 Å, was also observed in the
case of single-crystal gold films /12,13/. From this it is confirmed, that at least in the case of Cu and Au the conduction electrons are reflected mostly specularly at the <u>undisturbed</u> surfaces.
The scattering hypothesis based on (7) and (8) finds its real
justification here. Although the film surfaces are mainly oriented
in the direction of (111) and the Fermi surfaces, in that case,
do not show a spherical shape in any way /121/, the ordered surface of the single-crystal film does not contribute to film resistivity noticeably. Even the influence of lattice distortion
caused by the chemical bond between metal film and substrate or
by the roughness of the substrate, which could be propagated to
the film surface of the film/substrate interface, must be very
small.

(c) Numerous authors find in evaluating the thickness dependence of
the resistivity of evaporated metal films with the help of (4)
negative p-values /4,64,122,123/. Naturally no physical interpretation can be attached to negative p-values at the first instance.
But taking into account (18) this result may be traced back easily
to a marked roughness of the surface, leading to an effective
scattering cross section higher than the maximum value possible
in the case of plane parallel surfaces. At this point one gets
the first distinct experimental indication, that the scattering
hypothesis can be used in interpreting the thickness dependence
even in those cases, where the Fuchs-Sondheimer theory does not
lead to sensible results.

(d) At very low thicknesses resistivity values are observed, which
always lie distinctly above the theoretical curves. The critical
thickness, below which this effect is observed, depends not only
on film material, but also on the used substrate and on the degree
of vacuum. One can confirm this, for instance, by comparing the
results of fig. 20 with experimental data obtained by LAL and
DUGGAL /124/ for nickel films evaporated at a residual gas pressure of 10^{-5} Torr on mica substrates. These authors oberserved
a critical thickness value as high as 400 Å. In order to interpret
the abnormally high resistivities found for the thinner films
following models have been discussed in the literature:

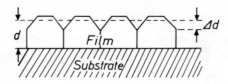

Fig.24 Fluctuation of film thickness in very thin films (schematically)

1. The band structure of the film material shall change effectively below a critical thickness because of the lattice distortion due to the interfaces, so that the number of free electrons decreases and the resistivity increases accordingly /125/.

2. Below the critical thickness the films may partly crack resulting in a formation of island and channels /40/.

3. The scattering parameter R (refer to (5)) shall increase with
   decreasing thickness, in certain cases by more than a factor
   of 5 /126/.

4. The local film thickness fluctuates remarkably about the analy-
   tically determined mean d-value due to the roughness always
   present on the surfaces (see fig. 24). Since the film regions
   with lower thicknesses shall contribute an extraordinarily
   high amount to the total resistivity, the measured ρ-values of
   the film will, therefore, be higher than expected from the
   mean thickness d.

Experimental evidence given by optical absorption and work
function measurements stands in contradiction to the first interpretation. Both quantities can be taken as a measure of the density of conduction electrons in the films /127/. The results obtained by RITZ /128/ and WÖLFING /61/ show that in the case of
nickel films no dependence of these quantities on film thickness
can be detected down to thicknesses of 40 Å. Changes are observed
only for films below 40 Å, which might be interpreted by a diminuation of electron density.

As far as the second interpretation is concerned an objection
may be raised, that a negative TCR should appear due to formation
of island structure /7/. Such TCR-values are observed, however, again
only below 40 Å for nickel films evaporated under ultrahigh vacuum

conditions (this behaviour may differ remarkably, if contanimation of the film is provided by evaporating at 2×10^{-5} Torr /135/).

The third interpretation is specially problematical. For, it is plainly to be seen, that each thickness dependence curve determined experimentally can be approximated by any theory as well, as one likes, if only a thickness dependence of the scattering parameter (R, p, ZA, $Z^* A^*$) is presumed and a proper choice of this dependence is guaranteed. In our case it seems to be diffi-cult to understand, why, for instance, the fraction R of electrons reflected specularly at the grain boundaries should vary with thickness.

Therefore, the fourth interpretation appears to be nearest to reality. The effect of a macroscopical roughness on the resisti-vity was first taken into consideration by WEDLER et al /11/ by introducing a so-called thickness correction term Δd. This term was assumed to be thickness independent and was determined to be about 25 Å in the case of nickel films. These results were con-firmed by evaluating stereo photographs obtained in an electron microscope, which led to roughnesses of the same order of magni-tude /79/. Later a more detailed treatment of the problem was given by NAMBA /102/, who computed the resistivity increase under the assumption of a sinusoidal film thickness distribution.

Finally it should be mentioned, that many theories /96,97,101/ predict a saturation effect of resistivity with decreasing film thickness. Such a saturation,however, has never been seen for the films discussed here. It may be concluded, therefore, that the prevailing roughness effect prevents a comprehensive experimental proof of these theories in most of the practical cases.

5. The Temperature Dependence of Resistivity of Pure Metal Films

The applicability of the different models developed for the description of resistivity behaviour of evaporated metal films can be checked by investigating other electrical properties on the same films additionally /6/. Temperature coefficient of resistivity (TCR), thermoelectric power (TP) and Hall effect (HE) can serve as some of them. In this section we shall deal with the temperature dependence of the resistivity, firstly. For quantitative calculations we shall use the so-called absolute temperature coefficient of resistivity (ATCR), which is defined by

$$\beta = d\rho/dT \tag{29}$$

This can be easily recalculated from the TCR-values α given normally in the literature, if the resistivity ρ is known, with the help of the relation

$$\alpha = \beta/\rho \tag{30}$$

5.1 Theoretical Aspects

For the sake of simplicity the calculation of the ATCR is started by using (4), (6), (12) or (15), respectively, which gives the tickness dependence of resistivity in the region under investigation with sufficient accuracy. Differentiating with respect to temperature one gets

$$\beta = \beta_o + \frac{3}{8}(1-p)\frac{1}{d}\frac{d(\rho_o l_o)}{dT} - \frac{3}{8}\frac{\rho_o l_o}{d}\frac{dp}{dT} \tag{31a}$$

(Fuchs-Sondheimer theory)

$$\beta = \beta_o + \frac{3}{2}\frac{1-R}{R}\frac{1}{D}\frac{d(\rho_o l_o)}{dT} + \frac{3}{2}\frac{\rho_o l_o}{D}\frac{d(1-R)/R}{dT} \tag{31b}$$

(Mayadas-Shatzkes theory)

$$\beta = \beta_o + \frac{z^* A^*}{D}\frac{d(\rho_o l_o)}{dT} + \frac{\rho_o l_o}{D}\frac{d(z^* A^*)}{dT} \tag{31c}$$

(Scattering hypothesis)

$$\beta = \beta_o + \frac{ZA}{d} \frac{d(\rho_o l_o)}{dT} + \frac{\rho_o l_o}{d} \frac{d(ZA)}{dT} \qquad (31d)$$

(Scattering hypothesis)

where β_o is the ATCR of the bulk.

According to the free electron gas model the product $\rho_o l_o$ is a measure of the density of conduction electrons and, therefore, does not depend on temperature to a first approximation /1/. Eqs. 31 then reduce to

$$\beta = \beta_o - \frac{\rho_o l_o}{d} \frac{3}{8} \frac{dp}{dT} \qquad (32a)$$

$$\beta = \beta_o + \frac{\rho_o l_o}{D} \frac{3}{2} \frac{d(1-R)/R}{dT} \qquad (32b)$$

$$\beta = \beta_o + \frac{\rho_o l_o}{D} \frac{d(Z^* A^*)}{dT} \qquad (32c)$$

$$\beta = \beta_o + \frac{\rho_o l_o}{d} \frac{d(ZA)}{dT} \qquad (32d)$$

The second term of right side of these equations characterizes in each case the contribution of conduction electrons, which are scattered diffusedly at lattice defects due to the surface or grain boundaries. If Matthiessen's rule is valid, then this contribution should also not depend on temperature, so that in all cases

$$\beta = \beta_o \qquad (33)$$

The ATCR should, therefore, take independent of thickness as well as annealing temperature, a β-value equal to bulk value. Calculating the TCR-value, the combination of (30) and (33) with (4) leads to

$$\alpha_o/\alpha = 1 + (3/8) \ (1-p) \ l_o/d \qquad (34a)$$

or transforming

$$d/\alpha = d/\alpha_o + (3/8) \ (1-p) \ l_o/\alpha_o \qquad (34b)$$

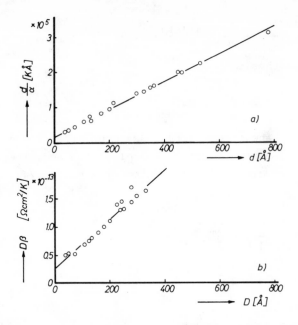

Fig.25 a) TCR for the nickel films of fig. 20 as a function of film thickness
(WÖLFING /61/). Plot in order to prove (34b)
b) Plot for the same films to prove (32c)

The relation (34) is commonly used in the literature for evaluation of
experimental data with the help of Fuchs-Sondheimer theory /1,4,6/.
Only a few papers exist, in which the validity of (33) is proved. The
reason for this may be, that in order to calculate the ATCR-values
the geometry factor F (refer to (1)) must be known. On the other hand,
taking into account (33) gives just an additional information, because
often an evaluation with the help of (34) leads to reasonable results
although (33) is not satisfied (compare, for instance, fig. 25 and
fig. 26). In the present section, therefore, we will emphasize the discussion of ATCR-values.

5.2 Experimental Results

First we shall report the results on the same nickel films for
which the resistivity behaviour has already been discussed in the
section 4.2 (refer to fig. 17). Since the evaluation of such measurement series has been done in the literature till now nearly exclusively with the help of Fuchs-Sondheimer theory, we have at first plotted the TCR-data in a diagramm having a linear scale each for ordina-

46

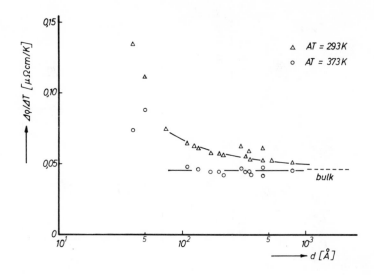

Fig.26 Δρ/ΔT-values for the films of fig. 25a as a function of film thickness.
Solid curves calculated from (32d)

te d/α and abscissa d. In such a plot the points should lie on a
straight line following (34b). It is recognized immediately from
fig. 25a, that this is fulfilled satisfactorily in the case of nickel
films annealed at 293 K. The characteristic constants computed from
slope and axis intercept are listed in table IV row 1. The result is
an (1-p) l_o-value of 55 Å for 293 K, which differs extremely from the
corresponding one determined from the resistivity values [l_o (1-p) =
350 Å, refer to table III]. Obviously the consequential application
of the Fuchs-Sondheimer theory without additional assumptions leads
to discrepancies. Nevertheless it must be seen, that these conclusions
depend strongly on the material of the films under investigation, and
on the deposition parameters. KINBARA et al /129/ reported, that in the
case of copper films a good consistency between both determination
methods was found. The non-applicability of the Fuchs theory in the
case of nickel films becomes more intelligible, when the thickness
dependence of the related ATCR is considered. Fig. 26 shows the measuring points of WÖLFING /61/ in a semi-logarithmic plot, where corresponding to fig. 20 the annealing temperatures are 293 K (Δ) and
373 K (o), respectively. One recognizes, that the constancy of ATCR
as required by (33) is realized only for AT = 373 K and d > 70 Å.
The values are, then, of the same order as the bulk. Below 70 Å the
ATCR-values lie, as in the case of resistivity measurements, too high

Table IV: Characteristic Constants of Thickness Dependence of TCR and ATCR of Evaporated Nickel Films

| AT* [K] | Reference | α_o [10^{-3} K^{-1}] | β_o [10^{-8} Ωcm K^{-1}] | evaluated with the help of equation | | |
				(32c) $\rho_o l_o\, d\,(Z^* A^*)/dT$ [10^{-14} Ωcm^2 K^{-1}]	(32d) $\rho_o l_o\, d\,(ZA)/dT$ [10^{-14} Ωcm^2 K^{-1}]	(34b) l_o for p=o [Å]
293	/61/	2,6	–	–	–	55
293	/61/	–	4,5	–	2,5	–
293	/61/	–	4,8	2,0	–	–
373	/61/	5,1	4,5	–	< 0,4	< 10
bulk	/113/	6,9	4,6	–	–	133

* AT stands for annealing temperature

(by a factor of about 2 for a 50 Å nickel film). Here again one should correlate the rise with the increasing roughness (refer to the discussion at the end of section 4.2). Rearranging (1) with the help of (33) we have

$$\Delta R/\Delta T = F \cdot f \, \beta_o/d \qquad (34)$$

so that the thinner film regions enter the average value with higher weight thus causing highly inflated ATCR-values /102/.

If one takes additionally the films annealed at 293 K into consideration, then the ATCR-values become dependent on thickness in the total investigated thickness region. That means, at least one of the assumptions necessary for the derivation of (33) is no longer fulfilled. We must conclude from our results, that Matthiessen's rule does no longer hold in rigorous sense. The phrase "in rigorous sense" is used to point at the fact, that the different resistivities compound together always additively to a total resistivity, but on other hand the resistivity component caused by scattering due to lattice defects becomes temperature dependent. Under these conditions (32) should be applied for quantitative evaluation instead of (33).

Setting $\rho_o l_o$ d (ZA)/dT equal to the positive value given in table IV column 6 leads to the curve plotted additionally in fig. 25. It is evident, that in this way the experimental points can be described satisfactorily. A similar result would be obtained by plotting βD against D according to

$$\beta D = \beta_o D + \rho_o l_o \, d \, (Z^* A^*)/dT \qquad (35)$$

which can easily be derived from (32c). Fig. 25b shows this plot for the same experimental values as given in fig. 25a. In this case too, by choosing the characteristic constants in a suitable manner (table IV column 5), a good conformity between theoretical and experimental values is found for d > 70 Å.

A further support for the correctness of this interpretation is to be seen in the fact that the effect of annealing temperature on ATCR can also be explained in the same way. The ATCR-values obtained for nickel films of approximately 200 Å thickness are plotted in fig. 22b against annealing temperature, where a drop is clearly seen. This drop

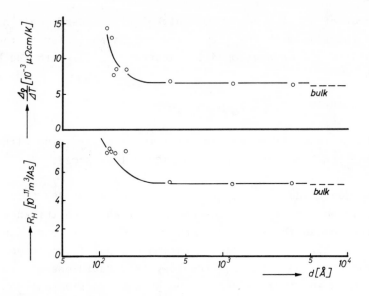

Fig.27 ATCR and Hall constant for copper films (MT = 273 K) as a function of
film thickness d (WEDLER and WIEBAUER /63a/)

can be described qualitatively by (32c) using numerical values from
table IV (solid curve in fig. 22b). Contrary to this falling curve, if
Matthiessen's rule were rigorously valid, then no dependence on annealing temperature is to be expected from (32).

It remains to explain, how a temperature dependence of the scattering parameter may occur. This effect is not restricted to nickel
films only, but has also been found by other authors working on alkali
metal /125/ and noble metal films /130,131/. For its interpretation
the following mechanisms might be considered:

(a) Tunneling of the electrons through a potential well between the
    crystallites: The formation of such potential wells is well-known
    in the case of ultra-thin metal films /133/ and semiconductor
    films /134/, and should result in a negative d (ZA)/dT /135/.

(b) Effect of thermally induced stresses: Completely annealed films
    are nearly free of stress /31,136/. After recooling, however,
    lattice distortions develop in the films due to different thermal
    expansion coefficients of film and substrate. These distortions,
    which cause a lattice contraction in the direction perpendicular
    to the film, can lead to an increase as well as to a decrease of

film resistivity /137-141/.

(c) Change in the phonon spectrum due to lattice disorder: In a simple
    picture one could imagine, that the uppermost atoms in the film
    surface or at the crystallite boundaries have, on the average, a
    larger spacing compared to the bulk because of missing neighbouring
    atoms. This effect is expected to be specially marked in the sur-
    roundings of a surface irregularity causing an increasing amplitu-
    de of vibrations /8/. This would lead to a positive d(ZA)/dT-value.

(d) Influence of anisotropic scattering: Since relaxation times of
    phonon and impurity scattering show different anisotropies, the
    resistivity component due to phonon scattering depends on the film
    thickness /142,143/. This effect can be calculated with the help
    of small-angle scattering models /143-145/. In the case of dilute
    copper alloys also experimental evidence exists, indicating that
    deviations from the validity of Matthiessen's rule may be contri-
    buted to such an anisotropic scattering /146/.

In the present work we found only positive d(ZA)/dT-values (refer
to table IV): The mechanism (a), therefore, may be excluded. Nevertheless, it is difficult to decide, which of the other mechanisms
preponderates. The good agreement of the experimental β_o-values of
the nickel films with the bulk value and the great difference of the
d(ZA)/dT-values determined for films annealed at 293 K and 373 K seem
to indicate that either the mechanisms (c) or (d) might be responsible
whereas (b) is highly unlikely.

Experimental values obtained for evaporated copper films of different thicknesses are plotted in fig. 27a (the corresponding resistivity values were given in fig. 23a). One finds a constant ATCR-
value for thicknesses greater than 200 Å, which deviates only very
little from the bulk value. Similar to the case of highly annealed
nickel films the ATCR-value rises only for thicknesses less than the
critical thickness defined above. Simultaneously an increase in the
Hall constant is observed (fig. 27b). We shall extend, therefore, our
discussion to include this quantity in the following.

5.3 Limits of the Free Electron Gas Model

We have seen in the preceding sections, that the resistivity behaviour and the temperature dependence of resistivity can be interpreted with astonishing successes with the help of the simple model
of free electron gas even in the case of nickel and copper films. The
two-band-structure of these transition metals, departures from the
spherical shape of the Fermi surface and anisotropic electron scattering need not be taken into consideration while discussing the effect
of thickness and annealing temperature dependence. One is tempted to
try to utilize the same simple model for explaining effects of more
complicated nature like Hall effect of thermoelectric power. The
attempt seems to be quite fruitful at the first sight, looking at
fig. 27b, which shows the thickness dependence of Hall constant R_H
(copper films, MT = AT = 273 K, measured by WIEBAUER /63/). Explanation of this dependence succeeds with the help of the scattering
hypothesis without any additional assumptions. Since the Hall constant
given by /1/

$$R_H = 1/eN \tag{37}$$

is a direct measure of the density N of free electrons, it is clear
that for higher thicknesses a constant R_H-value is arrived at, which
does not differ much from the bulk.

The rise in R_H for thicknesses below 200 $\overset{o}{A}$ can be traced back
again to the influence of roughness mentioned already at the end of
section 4.2. According to /63/

$$U_H \sim I \cdot B/d \tag{38}$$

where U_H is the Hall voltage, I the current through the film and B
the magnetic flux, U_H should increase inversely proportional to the
film thickness d. Therefore, similar to resistivity behaviour one gets
the effect, that the areas of lower thicknesses enter the average with
a disproportionately higher weight thus producing too high U_H and R_H
values.

It must be added, however, that, strictly speaking, we cannot come
to a decision on the basis of Hall effect measurements alone, whether
one or the other of the models presented in section 4.1 is to be preferred. For, it can be easily checked, that the thickness dependence
shown in fig. 27b may also be interpreted with the aid of FUCHS-SOND-

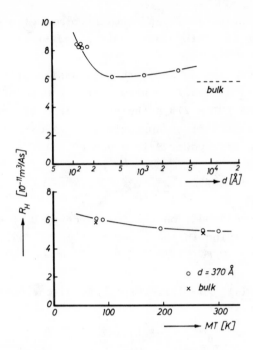

Fig.28 Effect of film thickness (MT = 77 K) and measuring temperature on the
Hall constant for copper films (WIEBAUER /63/)

HEIMER /10/ and MAYADAS-SHATZKES theory /86/ satisfactorily.

The good agreement between the experimentally obtained R_H-values
of fig. 27b and the theoretical curves computed from various theories
on the basis of the free electron gas model, is, however, more or less
accidental and has no general validity. This can be deduced from the
fact, that even for discussion of Hall effect measurements on bulk
copper the superposition of isotropic and anisotropic scattering components must be taken into account /115,116,145/. For copper films
this superposition can reveal itself in a Hall constant rising with
increasing thickness, as is demonstrated in fig. 28a (measurements
performed at MT = 77 K by WIEBAUER /63/). Furthermore the Hall constant shows a remarkable temperature dependence (fig. 28b), which
becomes larger for thinner films /147/. Both effects contradict the
predictions of scattering hypothesis as well as that of Fuchs-Sondheimer and Mayadas-Shatzkes theory, showing thereby the limitations
of the capacity of the used models. These limitations will be more
distinct in the case of nickel films, where carbon monoxide included

in the grain boundaries /148/ and the ferro-magnetic domain structure /149/ must be taken into consideration additionally.

Similar difficulties arise in the discussion of thermoelectric power measurements, as shall be shown for the case of nickel films in the following. At MT = 273 K the phonon drag effect is small and can be neglected in the first approximation /150/. Therefore, the absolute thermoelectric power S is given by /113/

$$S = C \; (\partial \ln\rho / \partial \ln E)_{E=E_F} \tag{39}$$

The constant C depends only on the Fermi energy E_F and on the absolute temperature, carrying in the present case (Ni-films, MT = 273 K) a value of C = 0,62 µV/K /57a/.

Setting for ρ (4) and (12), respectively, one gets after performing the differentiation

$$S \; (d) = S_o + C \left[1 + \frac{d}{0,41_o \, (1-p)} \right]^{-1} \left[\left(\frac{\partial \ln l_o}{\partial \ln E} \right)_{E_F} + \left(\frac{\partial \ln (1-p)}{\partial \ln E} \right)_{E_F} \right]$$

(Fuchs-Sondheimer theory) (40a)

$$S \; (d) = S_o + C \left[1 + \frac{D}{z^* \, A^* l_o} \right]^{-1} \left[\left(\frac{\partial \ln l_o}{\partial \ln E} \right)_{E_F} + \left(\frac{\partial \ln z^* \, A^*}{\partial \ln E} \right)_{E_F} \right]$$

(Scattering hypothesis) (40b)

with

$$S_o = C \left(\frac{\partial \ln \rho_o}{\partial \ln E} \right)_{E_F}$$

The thickness dependence of S as expected from the above formulae is presented in fig. 29. The solid curve was computed from (40a), the dashed curve from (40b) taking into account the experimentally obtained relation D = D (d) (refer to fig. 6a). The experimental points obtained by REICHENBERGER /57a/ fit excellently, if one chooses following values for the charcteristic constants:

$$l_o \; (1-p) = 300 \; \mathring{A} \quad ; \quad S_o = -16,5 \; µV/K \tag{41a}$$

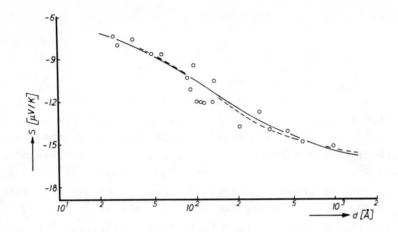

Fig.29 Absolute thermoelectric power S as a function of thickness of nickel films
(AT = 333 K, MT = 273 K) (WEDLER et al. /57a/)
—— computed from (40a)   --- computed from (40b)

$$\left(\frac{\partial \ln l_o}{\partial \ln E}\right)_{E_F} + \left(\frac{\partial \ln (1-p)}{\partial \ln E}\right)_{E_F} = 19 \qquad (41a)$$

and

$$z^* A^* l_o = 200 \text{ Å} \quad ; \quad S_o = -19,5 \text{ } \mu V/K$$

$$(41b)$$

$$\left(\frac{\partial \ln l_o}{\partial \ln E}\right)_{E_F} + \left(\frac{\partial \ln z^* A^*}{\partial \ln E}\right)_{E_F} = 23$$

respectively.

It is evident from fig. 29, that both the computed curves explain the experimental results satisfactorily, whereby the agreement for the dashed curve (computed with the help of scattering hypothesis) appears to be a little bit better. The S_o-values coincide with the bulk value (S_b = 19 $\mu V/K$ /151/) to a great extent, while the $l_o(1-p)$ and $z^* A^* l_o$ values are nearly equal to the corresponding values obtained from resistivity measurements on the same films (see table III).

Difficulties arise, however, while interpreting the high numerical values obtained for the sum

$$(\partial \ln l_O / \partial \ln E)_{E_F} + (\partial \ln Z^* A^* / \partial \ln E)_{E_F}$$

On the basis of free electron gas model numerical value of the first term of the sum should be only 2 /152/, the second term has to disappear under the assumption $Z^* A^* = $ const. Evidently, the real band structure of nickel must be taken into account in the present case, if a better agreement between theory and experiment is to be achieved. The discrepancies become smaller, if copper films are considered. GOUAULT /153/ determined for the above sum the value of 2,13 which does not differ noticeably from the theoretical prediction.

Finally it is interesting to note, that (40) for $d \ll 0.4\, l_O(1-p)$ and $D \ll Z^* A^* l_O$, respectively, goes over to a thickness independent form. In this way it becomes understandable, as to why the fluctuations of film thickness, which play a very important role in increasing the resistivity, TCR and Hall constant, do not influence the thermoelectric power. Fig. 29 demonstrates, that the experimental points lie on the theoretical curves down to thicknesses much below the critical thickness, which were found to be 70 $\overset{o}{A}$ in the case of nickel (refer to the end of section 4.2).

6. Resistivity Change Due to Gas Adsorption

So far we have considered only uncovered metal films, which had surfaces as clean as possible. Now we shall treat the change in electrical properties which occurs as a result of adsorption of gas on the films. Limiting the discussion at first to resistivity changes only, one finds experimentally the dependence shown schematically in fig. 30. The resistivity always increases proportionally to coverage during the first stages of adsorption. Presumption for this be-

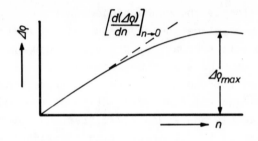

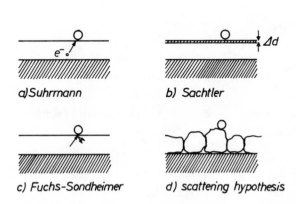

Fig.30 Resistivity increase due to gas adsorption and models suitable for interpretation (schematically)

haviour, however, is that the surface is not already poisoned with
certain amounts of foreign gas molecules and that the adsorbate gas
does not diffuse into the bulk like oxygen. At higher coverages deviations from the proportionality take place. The curve either achie-
ves a saturation value or runs through a maximum.

The proportionality constant $\left[d(\Delta\rho)/dn\right]_{n\to o}$ can serve as a characteristic quantity for the description of the measured curves. In the
following, therefore, we shall review the theoretical models proposed
in the literature for the calculation of this constant.

6.1   Theoretical Treatment

6.1.1 Suhrmann Model

The first model for the treatment of the above problem was put
foreward by SUHRMANN and co-workers /22-24/. These authors assumed

that a direct transition of the conduction electrons over to the
adsorbed molecules takes place thereby reducing the number of free
electrons in the metal (a schematical presentation of the mechanism
is shown in fig. 30). At the same time the mean free path l_o of the
electrons and the mean thickness d of the films were considered to
remain unchanged. The quantitative formulation of the model is described in detail by WEDLER and FOUAD /27/. Starting point is again
the free electron gas model. Differentiating logarithmically one
gets from (11) taking into account /1/

$$mv \sim N^{1/3} \qquad (42)$$

the relation

$$\Delta\rho/\rho = - (2/3) (\Delta N/N) \qquad (43)$$

Replacing the electron density N by the absolute number s of free
electrons leads to

$$\Delta\rho/\rho = - (2/3) (\Delta s/s) \qquad (44)$$

s increases proportional to the number of metal atoms and, therefore,
is related proportionally to the film thickness

$$s \sim d \qquad (45)$$

In addition to this, at first small gas doses, Δs should increase
proportionally to the coverage, because the electron transition is
directly coupled with the adsorption process and because the interaction between particles situated far from each other is negligible:

$$|\Delta s| \sim n \qquad (46)$$

Combining (44) with (45) and (46) and denoting the proportionality
constant by K_1, we have

$$\Delta\rho/\rho = K_1 n/d \qquad (47)$$

from which it follows

$$\left[d(\Delta\rho/\rho)/dn\right]_{n \to o} = K_1/d \qquad (48)$$

This model was attacked by other authors on the grounds, that adsorbed molecules should also be able to donate, in principle, conduction electrons to the metal. This would have resulted in a resisti-

vity decrease, which was never seen experimentally at low coverages
/25/. Further, rough calculations showed that the changes in the
electron density should remain restricted to the immediate vicinity
of the surface due to extremely small screening radii /8/. These
considerations led SACHTLER and co-workers /28/ to the assumption
that the uppermost atom layer of the metal lattice looses its metallic properties either partly or completely due to chemical binding
with the adsorbate thus no longer taking part in conductivity. A
consequence of the above would be a decrease of effective film thickness simultaneously leaving the electron density N and the mean free
path l_o more or less unchanged (schematically sketched in fig. 30).

6.1.2 Sachtler Model

For a quantitative formulation of the model one begins appropriately with the relation

$$\rho/\rho_o = f\ (d,AT) \tag{49}$$

where f (d,AT) is the experimentally determined dependence of the
resistivity on the deposition parameters film thickness d and annealing temperature AT, which might be approximated, for instance, by
(3). For a free electron gas we obtain by rearranging (11) and (42)

$$\rho_o \sim N^{-2/3}\ l_o^{-1} \tag{50}$$

therefore

$$\rho \sim \frac{f(d,AT)}{N^{2/3}\ l_o} \tag{51}$$

Differentiating logarithmically with respect to film thickness one
gets

$$\frac{1}{\rho}\frac{\Delta\rho}{\Delta d} = \frac{1}{f(d,AT)}\frac{\Delta f(d,AT)}{\Delta d} = h\ (d,AT)/d \tag{52}$$

The function f(d,AT) to be differentiated can be determined easily
from thickness dependence of resistivity measured at different annea-

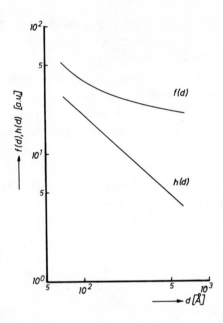

Fig.31   The quantities f and h of (52) as a function of film thickness (AT = 293 K, MT = 273 K)

ling temperatures (refer to fig. 20). For the case of nickel films annealed at room temperature fig. 31 shows the thickness dependence of f(d) obtained from fig. 20. Differentiating graphically one gets h(d), which is plotted additionally in fig. 31. The annealing dependence of the quantity h is illustrated by fig. 32 for a 200 Å thick nickel film.

Taking into account that the mean thickness decrease Δd for low coverages is again proportional to coverage, that is, setting

$$|\Delta d| \sim n \tag{53}$$

(52) may be written as

$$\Delta\rho/\rho = K_2 \; h \; (d,AT) \cdot n/d \tag{54}$$

where K_2 denotes the proportionality constant. Thus

$$\left[d(\Delta\rho/\rho)/dn\right]_{n\to o} = K_2 \; h \; (d,AT)/d \tag{55}$$

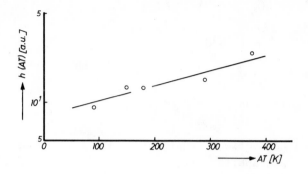

Fig.32 h as a function of annealing temperature (d = 200 Å, MT = 77 K)

It must be mentioned, that recently IONESCU /154/ has shown, that the demetallisation effect described by Sachtler may comprise a variation of conduction electrons to a certain degree. A quantitative calculation of this effect should be possible, if the simple proportionality between the number of free electrons s and the film thickness d (45) is taken into account.

6.1.3 Fuchs-Sondheimer Model

From the results obtained from resistivity measurements on single crystal films (refer to section 4.2) it should be expected, that the electrons are not reflected completely diffusedly but at least partially specularly at the film surfaces. This opens the possibility of interpreting the change in resistivity caused by gas adsorption as an additional scattering effect (schematically illustrated in fig. 30). If one is sure that the limiting surfaces of the film are plane parallel and do not have a noticeable roughness, then one can use (4) for a quantitative description /155,156/. Since according to this model gas adsorption results only in a decrease of the fraction p of the specularly reflected electrons, it follows from (4) by differentiation

$$\Delta\rho = -\frac{3}{8}\frac{\rho_o l_o}{d}\Delta p \qquad (56)$$

Setting analogous to (46) and (53)

$$|\Delta p| \sim n \qquad (57)$$

one gets

$$\Delta\rho = K_3 \; \rho_o l_o \; n/d \qquad\qquad (58)$$

and it follows

$$\left[d(\Delta\rho)/dn\right]_{n\to o} = K_3 \; \rho_o l_o/d \qquad\qquad (59)$$

A more detailed theoretical treatment of the problem was given by
WATANABE /157/, who applied Greene's boundary condition /158,100/
to adsorption effects in order to calculate the scattering cross
sections of the adsorbed molecules.

6.1.4 The Scattering Hypothesis

If the presumption of the plane-parallelness of the surface is
not fulfilled, then one should start from (15) instead of (4). In
this case the number Z of scattering centers per cm^2 of the film
surface is given directly by the coverage n

$$Z = n \qquad\qquad (60)$$

Putting it in (15) and rearranging one gets

$$\Delta\rho = \rho_o l_o \; A \; n/d \qquad\qquad (61)$$

which can be transformed into

$$\left[d(\Delta\rho)/dn\right]_{n\to o} = A \; \rho_o l_o/d \qquad\qquad (62)$$

a relation, which looks quite similar to (59). So we have at our dis-
posal four relations (48), (55), (59) and (62), which permit the calculation of the slope of the coverage curves as shown in fig. 30 in
the beginning stages of adsorption. Since the dependence of slope on
annealing temperature and film thickness is different in the different models, it becomes possible to check the validity of the above
formulae by comparing with experimental data, which is done in the
following section.

6.2   Experimental Results

6.2.1 Effect of Film Thickness

Fig. 33 shows the coverage dependence of the resistivity increase for four representative examples. The curves refer to metal/gas systems with weak bond (Ni/N$_2$ /60/ and Cu/CO /62,63/) as well as to systems with typical chemisorption properties (Ni/CO /57,27/ and Ni/H$_2$ /59,67/). All curves exhibit the typical linear shape at the beginning. Numerous examples of this behaviour are found in the literature too /3,8,159-162/. An important presupposition for the experimental proof of this linearity is, however, a high accuracy of coverage determination. Here the great advantage of gas adsorption measurements becomes evident in comparison with dielectric /163/ or metal adsorption experiments /17,164/.

To compare the thickness dependence calculated from the above mentioned models with experimental values we have plotted in fig. 34 the reciprocal thickness against the quotients $\left[d(\Delta\rho/\rho)/dn\right]_{n\to o}$ obtained from the coverage curves typically shown in fig. 33a and 33d. Apart from this, theoretical curves are plotted additionally, which were calculated under the assumption of either a change in the number of free electrons (curve I, computed from (48)) or a change in the effective film thickness (curve II, computed from (55)). In performing the calculation we have adjusted the values of at first unknown constants K_1 and K_2 to the experimental values of the high thickness range.   Fig. 34 shows clearly, that the experimental points do not fit any curve. It is evident, therefore, that the models forming the basis of these curves are not suitable for an explanation of the thickness dependence. (Similar conclusions regarding the assumptions leading to (48) were already drawn by other authors /27,157/). Further it is not understandable why the relative resistivity change for films of comparable thickness in the system Cu/CO is greater by a factor of about five than in the system Ni/CO, although the carbon monoxide is bound on nickel rather strongly as compared to copper /3/. On Sachtler model it were to be expected, that the relative resistivity change increases with increasing strength of the chemisorption bond /109/.

In order to decide whether (59) and (62) can describe the experimental results better, it is advisable to plot the <u>absolute</u> resisti-

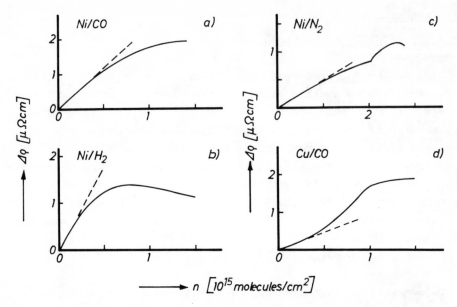

Fig.33 Dependence of the resistivity increase on coverage. Ni/CO (WENZEL /57/)
Ni/H$_2$ (REICHENBERGER /59/) Ni/N$_2$ (RICHTER /60/) and Cu/CO (KOCK /62/)

vity change $\Delta\rho$ against the film thickness instead of the relative
change $\Delta\rho/\rho$. Figs. 35a and 36b show the double-logarithmic plot for
the experimental points of fig. 34. The values obtained for the systems Ni/H$_2$ (fig. 35b) and Ni/N$_2$ (fig. 36a) are added. If (59)
and (62) are valid, then all the experimental points should lie in
this representation on straight lines with the slope minus one. Such
straight lines are plotted in figs. 35 and 36 additionally. One can
see, that experimental data fit the lines quite well. For the system
Ni/H$_2$ this has already been proved more than fifteen years ago, although a satisfactory explanation at that time could not be given
/67/. Measurements on the systems Au/CO /161/ and Fe/N$_2$ /166/ also
confirm the thickness dependence given by the (59) and (62). In all
cases, however, the experimentally determined $\left[d(\Delta\rho)/dn\right]_{n\to o}$-values
lie too high for thicknesses below the critical thickness (70 Å for
nickel), which must be traced back to the roughness effect already
discussed in detail at the end of section 4.2. A mechanism, which
takes into account the influence of adsorption on island-type films
/8,167/, should be effective only for d < 40 Å in the case of nickel.

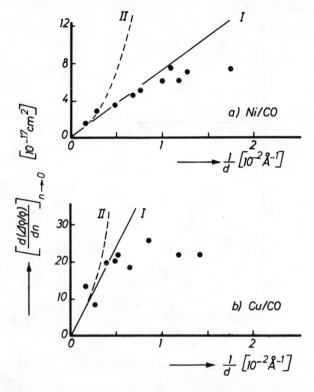

Fig.34 Relative increase of resistivity as a function of reciprocal thickness
a) Ni/CO (MT = 273 K) (WENZEL /57/)   b) Cu/CO (MT = 77 K) (KOCK /62/)
Curve I: Computed from (48) (Suhrmann model)
Curve II: Computed from (55) (Sachtler model)

Taking into consideration the real polycristalline structure of
the films under investigation (refer to fig. 8), the applicability
of (59) seems to be disputable. We shall restrict ourselves, therefore, in evaluating the experimental data only to (62). The numerical values of the characteristic constant $\rho_o l_o$ A obtained from the
axis intercept of fig. 35 and 36 are listed in table V, column 5.
These values are somewhat smaller than for a preceding evaluation /37/,
which can easily be traced back to the fact, that here the correction
factor f (refer to (1) and table I) is taken into consideration, while
in the previous paper /37/ it was omitted. From column 5, A-values
can be derived by putting in the $\rho_o l_o$-values from table III (i.e.
computed for the free electron gas model). The results which are
listed in column 6, show a reasonable order of magnitude but differ
from the value reported by WATANABE /157/ for the system Ni/H_2 (2 ×

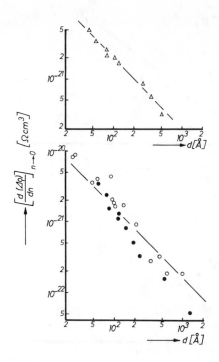

Fig.35 Experimental proof of (62) (scattering hypothesis)
 (Δ) Ni/CO AT = 333 K MT = 273 K (WENZEL /57/)
 (o) Ni/H_2 AT = 333 K MT = 273 K (REICHENBERGER /59/)
 (•) Ni/H_2 AT = 333 K MT = 77 K (REICHENBERGER /59/)

10^{-15} cm^2 at MT = 273 K). It is also easily seen from table V, that
the scattering cross section for the conduction electrons is smaller,
if the carbon monoxid is relatively weakly bound on copper, and is
larger, if it is chemisorbed on nickel. Obviously the paradox mentioned in the introduction concerning the correlation between resistivity increase and binding strength can be solved by considering the
absolute resistivity change instead of the relative.

Furthermore, it is very interesting to compare the resistivity
change due to gas adsorption with corresponding changes caused by
the inclusion of foreign atoms in the bulk of the metal. As is custo-
mary in the literature on diluted alloys it is advisible to refer
$\Delta\rho$ to $\bar{x}$ (that is atomic percent concentration of the alloyed foreign
atoms). From

$$n/d = \bar{N}\bar{x}/100 \qquad (63)$$

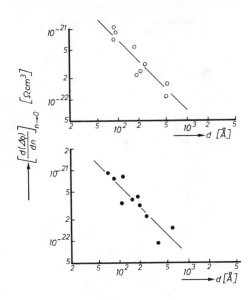

Fig.36 Experimental proof of (62) (scattering hypothesis)
        (o) Ni/N$_2$ AT = 293 K MT = 77 K (RICHTER /60/)
        (●) Cu/CO AT = 340 K MT = 77 K (KOCK /62/)

where $\bar{N}$ is the density of the excess metal atoms, and (61) it follows

$$\Delta\rho/\bar{x} = \bar{N} \, \rho_o l_o \, A/100 \qquad (64)$$

Setting in the known density value for nickel ($\bar{N} = 0,915 \cdot 10^{23}$ cm^{-3}) and the $\rho_o l_o$ A-value given in table V leads to

$$\Delta\rho/\bar{x} = 2,0 \ \mu\Omega cm/\% \ CO \qquad (65)$$

for the example of CO adsorption on nickel films. One recognizes, that this resistivity increase is of the same order of magnitude as that observed for nickel alloys /118/, which can be seen as a further indication of the correctness of the above interpretation.

6.2.2 Effect of Annealing Temperature

To check the validity region of (48), (55), (59) and (62), one can additionally take into consideration the dependence of initial slope of the coverage curves on the annealing temperature. In order to eliminate the effect of film thickness the discussion of the quan-

Table V: Scattering Cross Sections of Different Adsorbed Gases

System	Reference	AT [K]	MT [K]	$\rho_o l_o A$ $[10^{-27}\ \Omega cm^4]$	A $[10^{-16}\ cm^2]$
Ni/CO	/57/	333	273	2,2	2,6
Ni/H$_2$	/59/	333	273	1,8*	2,2*
Ni/N$_2$	/60/	293	77	0,7	0,9
Cu/CO	/62/	333	77	0,6	1,1

* In order to take into account the dissoziation of H$_2$ on a nickel surface /3/ these values are related to adsorbed hydrogen atoms.

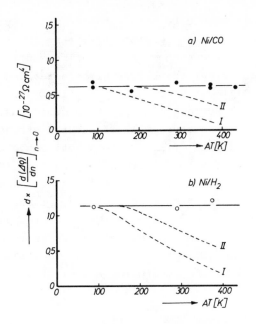

Fig.37 Resistivity increase as a function of annealing temperature (MT = 77 K, d ≈ 200 Å)
a) Ni/CO /58/   b) Ni/H$_2$ (evaluated from /61/ and /67/)
Curve I: computed from (48) (Suhrmann model)
Curve II: computed from (55) (Sachtler model)

tity d × $\left[d(\Delta\rho)/dn\right]_{n\to o}$ should be preferred in this case with respect to (62). Fig. 37 shows a plot of this quantity against annealing temperature. The experimental points stem from about 200 Å thick nickel films covered stepwise with CO /58/ and H$_2$ /67/, respectively. One finds, that within the experimental error no dependence of the resistivity increase on the annealing temperature is detectable. Theoretical curves computed from (48) and (55) have been added to the figure. In calculations using (48) (curve I) we have taken into account the function ρ = ρ (AT) determined for MT = 77 K experimentally by WÖLFING /61/, while for (55) (curve II) the dependence h = h (AT) was taken from fig. 32. There we have adjusted the at first unknown constants K$_1$ and K$_2$ to the experimental values for unannealed films. Evidently the experimental values fit neither curve I nor curve II. On the other hand, the measured independence of the resistivity increase on annealing temperature agrees very well with the prediction of (62). So these investigations lead again to the result, that the resistivity change due to gas adsorption may be correlated with an increased scattering of conduction electrons at the film surface.

Nevertheless it should be borne in mind while evaluating the annealing dependence with the help of (62), that the films possess in unannealed state a large number of highly indexed lattice planes at the surface, which orient themselves with increasing annealing temperature in the (111)-direction (refer to fig. 13). Since the orientation of the nickel surface changes it is quite possible to imagine, that the annealing temperature will affect the strength of interaction between gas and metal film due to a noticeable plane specificity thus influencing the scattering cross section A in (62). In the case of the system Ni/CO, however, this effect can be expected to be negligibly small, because measurements of heat of adsorption /168/ and surface potential /169/ indicate that with respect to the annealing process no plane specific adsorption can be found (Contrarily to this some plane specificity is observed, when (110)- and (100)-planes are involved in the adsorption /170-172/).

6.2.3 Effect of Measuring Temperature

If Matthiessen's rule is valid, then resistivity consists of a temperature dependent and a temperature independent fraction, where the first fraction is due to the electron scattering at the phonons and the second to scattering at lattice defects and foreign atoms /1/. Accordingly it should be expected, that on the basis of scattering hypothesis the molecules adsorbed at the surface should cause a temperature independent scattering /8/.

This prediction will be checked for the system Ni/CO in the following. This system is specially suited for a proof of the above hypothesis because the binding strength of carbon monoxide on nickel remains approximately the same irrespective of the temperature, at which it is adsorbed. To characterize the temperature dependence we may suitably define a quantity δ analogous to the absolute temperature coefficient of resistivity (29)

$$\delta = \frac{d \cdot \Delta \left[d(\Delta\rho)/dn \right]_{n \to o}}{\Delta T} \qquad (66)$$

Putting in (62) leads to

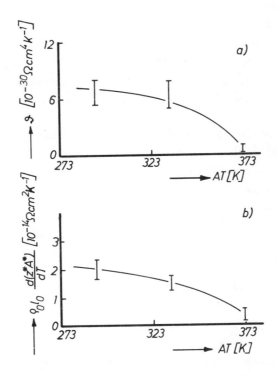

Fig.38 Temperature dependence of resistivity due to gas adsorption (a) and due to lattice disorder (b), plotted for comparison on the same abscissa

$$\delta = \frac{\left[\rho_o l_o A\right]_{T_2} - \left[\rho_o l_o A\right]_{T_1}}{T_2 - T_1} \tag{67}$$

Fig. 38a shows a plot of δ-values against the annealing temperature AT, where the measuring temperature is choosen to be $T_2 = 273$ K and $T_1 = 77$ K. The values were obtained by adsorbing CO stepwise on 200 Å thick nickel films /57,58,65/. It is evident, that for AT = 373 K the temperature dependence of the resistivity change due to gas adsorption disappears indeed. Therefore, under these conditions Matthiessen's rule is satisfied well. On the other hand, distinctly positive δ-values are observed for films annealed at lower temperatures. It is interesting to note, that this dependence on annealing temperature agrees strikingly well with the ATCR of "pure" films (fig. 38b, values taken from a previous paper /36/). This agreement favours the already proposed hypothesis, that the uncovered nickel films deposited

71

under ultrahigh-vacuum conditions still contain a noticeable amount of built-in carbon monoxide. Assuming, that the adsorbed and the built-in CO molecules possess the same scattering cross section A (which is a very rough approximation), one can estimate the amount of built-in CO on the basis of the measured resistivity increase due to adsorption. From fig. 38 this amount is determined to be $z^* \approx 3 \cdot 10^{15}$ molecules/cm^2, which corresponds to about two monolayers. Comparing this with the number of CO molecules striking the film surface during evaporation /3/ (evaporation time about 20 min), one comes to the conclusion, that all CO molecules arriving at the surface remain stuck. This result seems to be very reasonable.

A positive δ-value is also obtained for the hydrogen adsorption on nickel films, although with a remarkably weaker temperature dependence (fig. 35b). Similar results were found for binary alloys, for example for carbon steels /173/. The models based on changes in phonon spectrum or on influence of anisotropic scattering already discussed in section 5.2 can again be applied for interpreting these phenomena. We may point out here once more, however, that a positive δ-value for AT = 373 K cannot be observed in the Ni/CO system, although the roughness factor (refer to fig. 10) does not change much leading only to a relatively small change in the amount of adsorbed gas. Evidently, the influence of adsorbed CO overlayer depends on number of built-in lattice defects, which let us favour the anisotropic scattering model in explaining our results (see also section 6.5).

6.3 The Maximum Value of Resistivity Increase

Till now we have analysed only the linear part of the coverage curves. Fig. 33 showed, however, that the curves run through a maximum in many cases or tend to a saturation value. Under the presumption, that the film surfaces are plane parallel and have no noticeable roughness, one can calculate the maximum value with the help of the theory of Fuchs-Sondheimer. To start with, we take (56). Since p can take up values between 0 and 1 only, the resistivity increase remains restricted. The maximum Δp should be achieved exactly, if the scattering of conduction electrons at the interface film/vacuum takes place purely specularly before adsorption and completely diffusedly after adsorption. Since at the same time the scattering at the interface film/substrate remains unaffected, the maximum value is

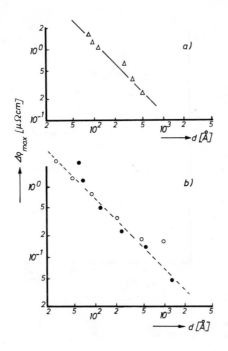

Fig.39 Experimental proof of (69)
        (Δ) Ni/CO AT = 333 K   MT = 273 K (WENZEL /57/)
        (o) Ni/H$_2$ AT = 333 K   MT = 273 K (REICHENBERGER /59/)
        ($\bullet$) Ni/H$_2$ AT = 333 K   MT = 77 K (REICHENBERGER /59/)

given by /38/

$$\Delta p_{max} = - 1/2 \tag{68}$$

The maximum possible resistivity increase is, then, given by

$$\Delta \rho_{max} = (3/16)\ \rho_o l_o/d \tag{69}$$

according to (56). To check (69) for all systems here under investigation the maximum increase in resistivity was determined from coverage curves (see fig. 30) and plotted on a double logarithmic scale
against thickness. Additionally straight lines with the slope of
minus one were drawn in the representation. It is seen from figs. 39
and 40, that the experimental points lie close to the straight lines,
that is the relation given by (69) is conformed as far as the thickness dependence is concerned.

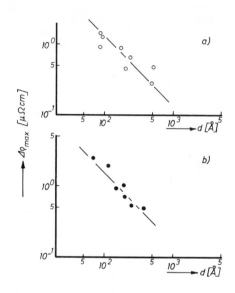

Fig.40 Experimental proof of (69)
          (o) Ni/N$_2$ AT = 293 K   MT = 77 K (RICHTER /60/)
          (•) Cu/CO   AT = 340 K   MT = 77 K (KOCK /62/)

The $\Delta\rho_{max}$-values read out for d = 100 $\overset{o}{A}$ are listed in table VI
column 4. Further this table contains theoretical values calculated
with the help of (69) and the $\rho_o l_o$-values taken from table III. The
comparison shows that the maximum resistivity increase calculated
on the basis of the free electron gas model agrees quite well with
the experimental values for the chemisorption system Ni/CO as well
as for the physisorption systems Ni/N$_2$ and Cu/CO. In spite of all
justifiable cautions, which should be borne in mind particularly in
the case of transition metals while evaluating quantitatively (69)
with the help of $\rho_o l_o$-values obtained from the simple free electron
gas model, this agreement appears to point at the fact, that the
conduction electrons are largely scattered specularly at the clean
film surface and only after a monolayer coverage are scattered completely diffusedly. That a sufficiently plane and clean metal surface does not contribute to diffuse scattering of the electrons - an
idea developed already in section 4 - finds here another important
experimental confirmation.

The system Ni/H$_2$ offers a certain exception from the general behaviour. Here, the experimentally determined resistivity increase
is smaller by a factor of 2 than the computed one. Evidently, the

Table VI: Maximum Resistivity Increase Due to Gas Adsorption

System	Reference	AT [K]	$\Delta\rho_{max}$ [$\mu\Omega cm$] for d = 100 Å	
			exp.	theor.
Ni/CO	/57/	333	1,25	1,4
Ni/H$_2$	/59/	333	0,70	1,4
Ni/N$_2$	/60/	293	1,45	1,4
Cu/CO	/62/	333	1,55	1,2
Ni/CO	/58/	90	8,10	1,4
Cu/CO	/70/	90	25	1,2

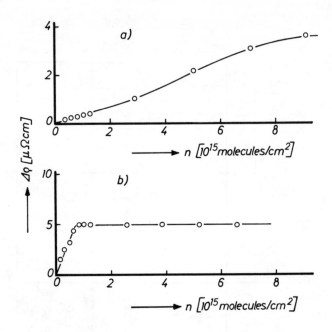

Fig.41 Resistivity increase of unannealed films during gas adsorption (MT = 77 K)
a) Ni/CO /58/ b) Cu/CO /70/

electrons are still partially specularly scattered just at a monolayer coverage. It could be imagined, that a nickel hydride phase
/174,175/ develops in the uppermost atom layer, which is not very
much different from pure nickel as fas as electrical properties are
concerned.

According to the free electron gas model the product $\rho_o l_o$ is a
measure of the number of free electrons (50) and is, therefore, independent of temperature. For the same reason according to (69) the
maximum resistivity increase should be independent of measuring temperature. This prediction can be easily checked for the system Ni/H$_2$,
if one compares the $\Delta\rho_{max}$-values determined by REICHENBERGER /59/ at
77 K and 273 K. One can see from fig. 39b immediately, that a dependence of the maximum resistivity change on measuring temperature
cannot be detected indeed.

In applying (69) it should always be assured, that the films are
free of roughnesses. The discussion is, therefore, limited only to
films annealed at relatively high temperatures (refer to fig. 8). At
very rough surfaces the area available to adsorption becomes relati-

vely large, and accordingly the maximum value of resistivity increase
given by (69) can be exceeded remarkably. A typical example is shown
in fig. 41, where a nickel and a copper film, both annealed at 90 K,
serve as adsorbent. It is evident, that in each case the maximum resistivity $\Delta\rho_{max}$ lies by a factor more than five higher than that given
in table VI, which was computed by using (62). The advantage of scattering hypothesis in comparison to the Fuchs-Sondheimer theory here
becomes specially explicit. Contrarily to (4), (15) has no upper limit for the resistivity increase, so that the curves presented in
fig. 41 can also be interpreted without any difficulty.

6.4 The Discussion of the Over-all Curve $\Delta\rho(n)$

Deviations from the initial straight line behaviour of the cove-
rage curves $\Delta\rho = \Delta\rho(n)$ can be computed, if one starts with the above
assumptions that the adsorbed molecules act as scattering centers
just as foreign atoms inside the bulk. The residual resistivity of
disordered binary alloys of different compositions follows the so-
called Nordheim's rule in several cases quite well /176/

$$\Delta\rho \sim x \; (1-x) \tag{70}$$

where x is mole fraction of one alloying component. (70) can be
immediately applied to the adsorption problems discussed here, if
the mole fraction is replaced by the coverage n as follows (refer
to (63))

$$x = n/\bar{N}d \tag{71}$$

denoting the density of metal atoms by $\bar{N}$. Setting in (71) leads to

$$\Delta\rho \sim (n/\bar{N}d) \left[1 - n/\bar{N}d\right] \tag{72}$$

A parabolic dependence on coverages is thus obtained, from which
the maximum resistivity change $\Delta\rho_{max}$ can be derived to lie at

$$n_{max} = \bar{N}d/2 \tag{73}$$

Putting this in (72) gives

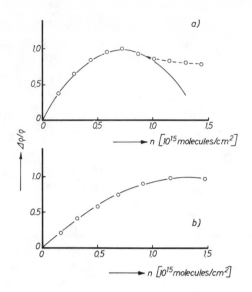

Fig.42 Coverage dependence of the resistivity increase
         a) Ni/H$_2$ MT = 273 K (REICHENBERGER /59/)
         b) Ni/CO MT = 273 K (WENZEL /57/)
         Solid curves calculated from (74)

$$\frac{\Delta\rho}{\Delta\rho_{max}} = \frac{n}{n_{max}} \; (2 - \frac{n}{n_{max}}) \qquad\qquad (74)$$

If, therefore, $\Delta\rho/\Delta\rho_{max}$ is plotted against n/n_{max}, then the experimental points should lie on one and the same parabola independent
of the respective deposition parameters like thickness, annealing
temperature etc. Fig. 42 shows typical curves obtained by REICHEN-
BERGER /59/ and WENZEL /57/ for H$_2$ and CO adsorption on nickel films.
It is evident, that experimental points fit the general course of
the parabolas quite well. Deviations are only observed at high coverage, where new adsorption states of weak binding energy (perhaps
the beginning of a second overlayer adsorption /65c/) can no longer
be excluded. It is possible, therefore, for the systems mentioned
above, to find out the initial slope from coverage curves by determining the maximum values $\Delta\rho_{max}$ and n_{max} and by using the relation

$$\left[d(\Delta\rho)/dn\right]_{n\to o} = 2 \; \Delta\rho_{max}/n_{max} \qquad\qquad (75)$$

which is obtained by differentiating (74). For an experimental check
of (75) the numerical values taken from the initial slope and the

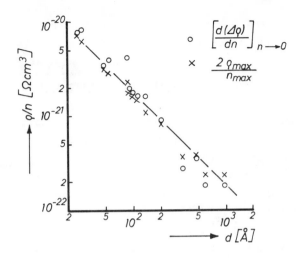

Fig.43 Experimental proof of (75)
        (points obtained by REICHENBERGER /59/ on Ni/H$_2$ system)

maximum position are compared in fig. 43. The experimental points
obtained by REICHENBERGER /59/ on the system Ni/H$_2$ confirm the relation in a very clear-cut manner.

   It should be noted, that during formation of a hydride phase in
bulk nickel a cross-over through a maximum resistivity has been observed to occur, too, for a composition H/Ni ≈ 0,8 /177/. In that
case, however, very high hydrogen pressures had to be used thereby
leaving the amount of built-in hydrogen indeterminably, so that a
calculation of the scattering cross section responsible for the resistivity increase could not be made.

   Contrary to the chemisorption systems (i.e. systems with relatively high binding energy as Ni/CO or Ni/H$_2$) deviations from the
parabolic curve shape arise in the case of typical physisorption
systems like Ni/N$_2$ or Cu/CO. In such systems a complete desorption
happens due to the weak nature of the bond between adsorbate and
adsorbent, if the film is brought to room temperature. Further
examples of such deviations may be found in numerous other papers
/3,8/. This is not surprising, however, because it is known, that
such deviations exist also amongst binary alloys. For example, a
curve shape similar to that shown in fig. 33d is observed for the
case of Ag-Pd-alloys /178/, while discontinuities in the coverage

curves (fig. 32c, at about $n = 2 \cdot 10^{15}$ molecules/cm^2) may point at formation of ordered phases /112/.

6.5 Changes in Hall Constant and Thermoelectric Power

The models so successfully explaining the resistivity measurements may also be attempted for checking whether an interpretation of the influence of gas adsorption on Hall constant and thermoelectric power is possible. Justification for this may be the useful informations already obtained for the case of clean films in section 5.3. If the scattering hypothesis (37) is valid, then the Hall constant should not depend on coverage. For any decrement in the number of free electrons or for a decrease of the effective film thickness /179, 26/ the Hall constant should be expected to increase with increasing coverage. Experimentally this behaviour has been found in the system Ni/CO /156/. On the other hand, a decrease is observed in the system Cu/CO /52,63/, which furthermore depends strongly on film thickness (fig. 44).

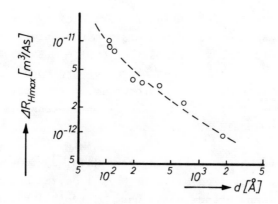

Fig.44 Maximum Change of Hall constant due to gas adsorption as a function of film thickness. (points obtained by WIEBAUER /63/ on Cu/CO system)

These results show once again, that the free electron gas model is not sufficient for the interpretation of Hall constants measured

on copper films (refer to the discussion in section 5.3). It should
rather be borne in mind, that different scattering mechanisms super-
pose in reality. This has been discussed very extensively by WIEBAUER
/63/, who advocates the view point that the anisotropic electron
scattering overweights in the films investigated at 77 K. Scattering
due to adsorbed molecules, which might be considered as effectively
representing a point defect, adds an isotropic scattering component
resulting in a decrease of the R_H-value towards the value determined
for a clean film at room temperature. Since the isotropic scattering
is effective only at the film surfaces, its influence should decrease
with increasing thickness, a prediction fully in agreement with the
experimental observation given by fig. 44.

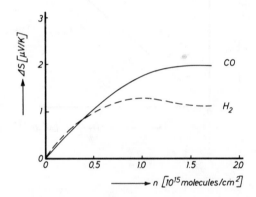

Fig.45 Effect of gas adsorption of the thermoelectric power of nickel films.
        MT = 273 K. Experimental curves obtained by WENZEL /57/ (—) and REICHEN-
        BERGER /59/ (--)

The same difficulties as already discussed in the case of pure
films have to be faced while interpreting the coverage dependence of
thermoelectric power S with the help of free electron gas model.
Fig. 45 shows the changes in S found for about 100 Å thick nickel
films subjected to CO and H_2 adsorption /36/. Thickness dependence
of the initial slopes of the coverage curves of fig. 45 is plotted
in fig. 46[*]. The large fluctuation of experimental values is quite

---

[*] To increase the precision of measurements the relation /36/

$$[d(\Delta S)/dn]_{n \to 0} = 2 \Delta S_{max}/n_{max}$$

was used for the evaluation of the hydrogen coverage curves. This
relation, equivalent to (75), was previously proved to be obeyed
by experimental data.

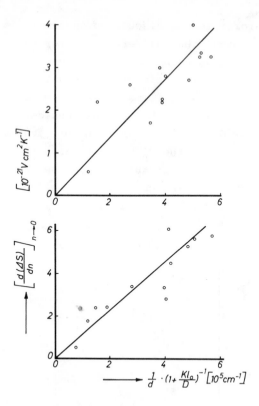

Fig.46 Thickness dependence of the change of thermoelectric power due to gas
adsorption
a) Ni/CO (WENZEL /57/) b) Ni/H$_2$ (REICHENBERGER /59/)
Plot according to (77)

conspicuous, in spite of which it has been attempted to explain the
thickness dependence with the help of Fuchs-Sondheimer theory and
scattering hypothesis even in this case.

Starting point for the calculation is again (39). Setting in
(17) and (60) leads to

$$S = S_o + C \left[\partial \ln (1 + z^* A^* l_o/D + n A l_o/d)/\partial \ln E \right]_{E_F} \qquad (76)$$

If it is taken into account, that the extra resistivity caused by
gas adsorption is small compared to the resistivity of uncovered
films, then after differentiation with respect to n we have

$$d(\Delta S)/dn = CAl_o \left[d (1 + z^* A^* l_o/D) \right]^{-1} \left[\partial \ln (Al_o)/\partial \ln E \right]_{E_F} \qquad (77)$$

In the plot of fig. 46, therefore, straight lines should be expected. If, in spite of the evident fluctuation in experimental values, such straight lines are drawn, then one can read out from the slope

$$\left(\frac{\partial \ln l_o}{\partial \ln E}\right)_{E_F} + \left(\frac{\partial \ln A}{\partial \ln E}\right)_{E_F} = 30 \qquad (78a)$$

in the case of CO adsorption and

$$\left(\frac{\partial \ln l_o}{\partial \ln E}\right)_{E_F} + \left(\frac{\partial \ln A}{\partial \ln E}\right)_{E_F} = 63 \qquad (78b)$$

in the case of H_2 adsorption, provided one takes into account the numerical values $C = 0,62$ μV/K /57a/, $l_o = 133$ Å (table III) and $A = 2,2 \cdot 10^{-16}$ cm^2 (table V). A comparison with (41) shows, that the values of (78) lie still higher than those obtained for uncovered nickel films. Thus, in this case too, the value 2 calculated on the basis of free electron gas model /152/ is exceeded by more than an order of magnitude.

7. Resistivity and Heat of Adsorption

Formerly resistivity measurements performed in connection with adsorption investigations served mainly for either determination of monomolecular coverage /3/, or detection of differently adsorbed species /180,181/, or adsorption mobility /27/ and kinetic studies /55/, or indication of reactions in the adsorption phase /54/. The discussion was always limited to a qualitative evaluation of the coverage dependence.

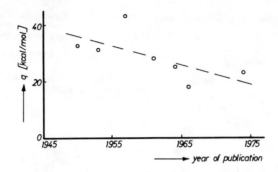

Fig.47 Plot to illustrate the influence of developing vacuum technique on the
experimental data /68a,182-187/ available in the literature for the heat
of hydrogen adsorption on nickel surfaces

In section 6 it was shown that a characteristic quantity, the mean
scattering cross section A, can be read out from these curves for a
quantitative analysis. The question arises now, what are the informations contained in the quantity A with respect to other adsorption
phenomena, and specially, whether any statements could be deduced
from it about the strength and sort of adsorption bond. Different
authors /3,37,109/ have attempted, for example, to bring the resistivity increase in relation with the heat of adsorption without being
able to give an unique picture.

Although the resistivity change and the heat of adsorption are related to each other definitely through the strength of electronic interaction between metal film and the adsorbed gas, an experimental
proof of this relationship is very troublesome. The problems of such
a proof may be seen in the fact, that the geometry factor of the experimental set-up necessary to compute the A-values is known with
sufficient accuracy only in very few cases. Moreover, the values of
the heat of adsorption given in the literature often fluctuate sub-
stantially, which can be traced back in many cases to the uncomparable
vacuum conditions under which the respective experiments were performed (A good example for this outcome is shown in fig. 47, where the
decrease of heat of hydrogen adsorption on nickel films seems to depend on the year in which it was measured /68a,182-187/). Further,
before undertaking any correlation of such experimental values, it
must be assured, that the experimental conditions and the deposition
parameters like film thickness, annealing and measuring temperature

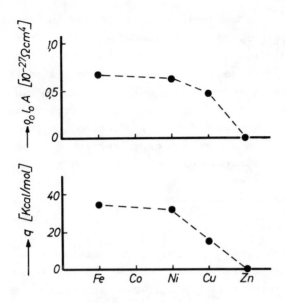

Fig.48 Plot to illustrate the correlation between resistivity increase and heat of adsorption. Details are given in the text

are in complete agreement.

For these reasons it is possible to compare within the framework of this paper only experiments on four different transition metals covered with CO. Fig. 48 shows the A-values as well as the ΔU-values in relation to the position of adsorbent metals in periodic table. The experimental values for resistivity increase were taken from RICHTER /60/ (Fe and Ni), KOCK /62/ (Cu) and SUHRMANN et al. /29/ (Zn), the heat of adsorption was measured under comparable experimental conditions /66,188-189/. From the figure one can see, that for iron and nickel films the cross section as well as the heat of adsorption take up specially high values. Evidently, both quantities are decidedly affected by the strong interaction of the carbon monoxide with the holes in the d-band. Going over to the copper films where d-holes are more or less occupied one finds smaller A and q values. Finally, in the case of Zn no adsorption is detectable at all.

The result of fig. 48 seems to indicate, that in certain cases the resistivity measurements may provide information about the strenght of adsorption bond at least at smaller coverages. How cautious one should be, on the other hand, while comparing the adsorption behaviour of different systems, is easily recognized considering the hydrogen adsorption on nickel. The hydrogen dissociates at the film surface /3/, which should cause an increase of effective scattering cross section A. At the same time it should be borne in mind that the heat of adsorption will be smaller because the energy necessary for dissociation must be supplied for. This discrepancy is a natural result of the fact that the scattering cross section A is solely determined by interaction of adsorbed gas molecules with the adsorbent, while the heat of adsorption is influenced additively by the mutual interaction of the adsorbed molecules.

It may be pointed out, that in contradiction to earlier attempts of interpretation /3/ no conclusive statements can be made on a prevailing covalent or polar character of the adsorption bond only on the basis of resistivity measurements. The increase in resistivity indicates a formation of new scattering centers, but it remains open to discussion, what form of special interaction mechanism is responsible for the disorder at the surface of the metal lattice.

8. Concluding Remarks

Aim of the present paper was to show that the resistivity behaviour of pure and gas covered films can be interpreted with the help of a simple scattering hypothesis. Only the free electron gas model forms the basis of this theory, and it is surprising, therefore, to find that within the limits of this simple approximation all effects observed just in the case of transition metals like nickel and copper can be explained satisfactorily.

Comparison of the experimental data obtained for these metals points to the fact that the complicated shape of the Fermi surface and the band overlap do not lead to any noticeable deviation, unless a similar effect exists for the bulk metal. Surprisingly this model can even explain the experimental data obtained for TCR, Hall constant and thermoelectric power within certain limits. On the other hand, evident discrepancies specially with respect to temperature dependence are found, which bring forth the limits of the highly simplifying assumptions underlying the model.

On the basis of the results discussed above the resistivity behaviour may be summarized as follows: Pure, ideal films show, independent of annealing temperature and thickness, the electrical properties of the corresponding bulk material. Only below a critical thickness, which is much lower for nickel (70 $\overset{o}{A}$) than for copper films (200 $\overset{o}{A}$), the effect of roughness becomes so large, that resistivity, TCR and Hall constant rise abruptly. The scattering of the conduction electrons at the scattering centers which are caused by the lattice defects in the bulk and at the film surface or by the built-in and adsorbed foreign molecules, manifest itself in rising resistivity too. Now, however, the resistivity increase remains effective at higher thicknesses also and can be altered by annealing treatment. Such a behaviour is to be expected, for instance, when either the films have a polycrystalline structure, or when the interface film/substrate acts as a noticeable distortion of the film surface.

Finally a few points shall be accentuated, which are important for a comparison with results of the published literature:

(a) Even for physisorbed, that is relatively weakly bound gas a single adsorbed monolayer is enough to change the Fuchs specularity p from one to zero for a film with plane surface (refer to the results obtained in sections 6.3 for the systems Ni/N_2 and Cu/CO. Similar results have been found, for example, for the system Au/CO /161/). Now, it is difficult to avoid such a physisorption specially in the case of low temperature measurements, where the films themselves sorb strongly. Therefore, the results of such investigations /15,20,191/ can only be brought in correlation with measurements on clean films, if the exact experimental conditions are known. The same is valid for galvanomagnetic measurements carried out normally at 4.2 K /4,192,193/.

(b) Similar reservations should be borne in mind while discussing results obtained at low temperatures on wires, foils and whiskers /5,143,145,194,195/. Here, too, the procedure used for prepara-

tion and cleaning the probe determines whether a gas-free surface
is provided or not.

(c) If the surfaces are really free of overlayers, then the conduction electrons shall be scattered more or less specularly even
in the case of polycrystalline films (refer to the results of
section 6.4). Simultaneously the films show tendency to the preferential formation of cubic crystallites, whose thickness is
approximately equal to the film thickness. Then, the resistivity
behaviour of the films is guided by the crystallite boundary
scattering leading to a thickness dependence which cannot be differentiated practically from that resulting from surface scattering. In this way one can understand, for example, the discrepancy
mentioned already in the introduction, that the interpretation of
the thickness dependence found experimentally seems to require
$p = o$, although optical measurements uniquely point to $p = 1$
simultaneously /21/.

(d) Tendency of the films towards the formation of cubic crystallites
is, in our opinion, the actual reason for the increasing influence
of the roughness observed for very thin films. This "macroscopic"
roughness leads to a marked decrease in the effective film thickness (refer to fig. 24), thereby hindering the use of the relations
derived in this paper, though the films are still intact, that
is they show no island structure. The critical thickness, where
this effect becomes noticeable, depends on the system under investigation. In general one finds $d_{crit.} (Ni) < d_{crit.} (Cu)$
$< d_{crit.} (Au) < d_{crit.} (Ag)$ /75/, where, however, the individual
numerical values are influenced by substrate and the vacuum conditions during deposition. In this context it may be recalled,
that nickel films deposited on mica at 10^{-5} Torr were shown to
behave abnormally even below 400 Å /124/.

I would like to thank Prof. Dr. G. Wedler for drawing my attention
to this field and for many helping discussions. Financial support by
Deutsche Forschungsgemeinschaft is gratefully acknowledged.

References

1. H. MAYER: Physik dünner Schichten. Stuttgart: Wissenschaftliche Verlagsgesellschaft, 1955, Band II, p. 178 ff.

2. A.C. TICKLE: Thin Film Transistors. New York: John Wiley Sons Inc., 1969, p. 121 ff.

3. G. WEDLER: Adsorption. Weinheim/Bergstraße: Verlag Chemie, 1970, p. 120 ff.

4. K.L. CHOPRA: Thin Film Phenomena. New York: Mc-Graw-Hill Book Co., 1969, p. 344 ff.

5. D.C. LARSON: in G. Hass and R.E. Thun (eds): Physics of Thin Films. New York: Academic Press, 1971, p. 81 ff.

6. T.J. COUTTS: Thin Solid Films $\underline{7}$, 77 (1971).

7. T.J. COUTTS: Electrical Conduction in Thin Metal Films. Amsterdam: Elsevier Publ. Co., 1974, p. 149 ff.

8. J.W. GEUS: in J.R. Anderson (ed): Chemisorption and Reactions on Metallic Films. London: Academic Press, 1971, Vol. 1, p. 129 ff.

9. K. FUCHS: Proc. Camb. Phil. Soc. $\underline{34}$, 100 (1938).

10. E.H. SONDHEIMER: Adv. Physics $\underline{1}$, 1 (1952).

11. G. WEDLER, F.J. BRÖCKER, H.G. KOCK and C. WÖLFING: in H. Mayer and R. Niedermayer (eds): Grundprobleme der Physik dünner Schichten. Proc. Intern. Symposium Clausthal-Göttingen 1964. Göttingen: Vandenhoeck und Ruprecht, 1965, p. 566.

12. K.L. CHOPRA, L.C. BOBB and M.H. FRANCOMBE: J. Appl. Phys. $\underline{34}$, 1699 (1963).

13. K.L. CHOPRA and L S. BOBB: Acta Metallurgica $\underline{12}$, 807 (1964).

14. P. BROQUET et V. NGUYEN Van: Surface Sci. $\underline{6}$, 98 (1967).

15. H.G. KADEREIT: Thin Solid Films $\underline{1}$, 109 (1967).

16. E.J. GILHAM, J.S. PRESTON and B.E. WILLIAMS: Phil. Mag. $\underline{46}$, 1051 (1955).

17. M.S.P. LUCAS: Thin Solid Films $\underline{2}$, 337 (1968).

18. H. HOFFMANN: remarks in discussion, see reference 11, p. 559.

19. J. CHAUVINEAU and P. CROCE: Compt. Rend. B $\underline{266}$, 1622 (1968).

20. A.F. MAYADAS, R. FEDER and R. ROSENBERG: J. Vac. Sci. Technol. $\underline{6}$, 690 (1969).

21. H.E. BENNETT and J.E. BENNET: in F. Abeles (ed): Optical Properties and Electronic Structure of Metals and Alloys. Amsterdam: North-Holland Publ. Co., 1966, p. 175

22. R. SUHRMANN: Ber. Bunsenges. Phys. Chem. 60, 804 (1956).

23. R. SUHRMANN and K. SCHULZ: Z. Phys. Chem. (Frankfurt) 1, 69 (1954).

24. R. SUHRMANN and G. WEDLER: remarks in discussion in: Ber. Bunsenges. Phys. Chem. 63, 757 (1959).

25. W.M.H. SACHTLER: remarks in discussion in: Ber. Bunsenges. Phys. Chem. 63, 757 (1959).

26. W.M.H. SACHTLER: Surface Sci. 22, 468 (1970).

27. G. WEDLER and M. FOUAD: Z. Phys. Chem. (Frankfurt) 40, 12 (1964).

28. W.M.H. SACHTLER and G.J.H. DORGELO: Z. Phys. Chem. (Frankfurt) 25, 69 (1960).

29. R. SUHRMANN, H. OBER and G. WEDLER: Z. Phys. Chem. (Frankfurt) 29, 305 (1961).

30. G. WEDLER and P. WISSMANN: Z. Naturforschung 23a, 1537 (1968).

31. G. WEDLER and P. WISSMANN: Z. Naturforschung 23a, 1544 (1968).

32. P. WISSMANN: Thin Solid Films 5, 329 (1970).

33. P. WISSMANN: Z. Phys. Chem. (Frankfurt) 71, 394 (1970).

34. P. WISSMANN: Thin Solid Films 6, R 67 (1970).

35. G. WEDLER, C. WÖLFING and P. WISSMANN: Surface Sci. 24, 302 (1971).

36. H. REICHENBERGER, G. WEDLER, H. WENZEL, P. WISSMANN and C. WÖFLING: Ber. Bunsenges. Phys. Chem. 75, 1033 (1971).

37. G. WEDLER and P. WISSMANN: Surface Sci. 26, 389 (1971).

38. P. WISSMANN: Thin Solid Films 13, 189 (1972).

39. P. WISSMANN and Y. SUMMA: Vakuum-Technik 22, 116 (1973).

40. C.A. NEUGEBAUER and M.B. WEBB: J. Appl. Phys. 33, 74 (1962).

41. R.M. HILL: Proc. Roy. Soc. A 309, 377 (1969).

42. K.L. CHOPRA: Phys. Rev. 155, 660 (1967).

43. Te-Chang LI and V.A. MARSOCCI: Phys. Rev. B 6, 392 (1972).

44. G.E. JURAS: Phys. Rev. B 2, 2869 (1970).

45. P. COTTI: Phys. Letters 4, 114 (1963).

46. R.A. POWELL, A.F. CLARK and F.R. FICKETT: Phys. Kondens. Mat. 9, 104 (1969).

47. T.T. CHEN and V.A. MARSOCCI: J. Appl. Phys. 43, 1554 (1972).

48. C. REALE: J. Material Sci. 6, 33 (1971).

49. G. BLIZNAKOV and D. LAZAROV: Z. Phys. Chem. (Leipzig) $\underline{223}$, 33 (1963).

50. J.W. GEUS, L.T. KOKS and P. ZWIETERING: J. Catalysis $\underline{2}$, 274 (1963).

51. I.G. MURGULESCU and N.I. IONESCU: Thin Solid Films $\underline{7}$, 355 (1971).

52. G. COMSA: Thin Solid Films $\underline{4}$, 1 (1969).

53. V. PONEC, Z. KNOR and S. CERNY: J. Catalysis $\underline{4}$, 485 (1965).

54. R. SUHRMANN, J.M. HERAS, L.V. HERAS and G. WEDLER: Ber. Bunsenges. Phys. Chem. $\underline{68}$, 511 (1964).

55. R. SUHRMANN, H.J. BUSSE and G. WEDLER: Z. Phys. Chem. (Frankfurt) $\underline{47}$, 1 (1965).

56. L.H. GERMER, I.W. MAY and R.I. SZOSTAK: Surface Sci. $\underline{7}$, 430 (1967).

57. H. WENZEL: Dissertation TU Hannover 1968

   a) G. WEDLER, H. REICHENBERGER and H. WENZEL: Z. Naturforschung $\underline{26a}$, 1444 (1971)

   b) G. WEDLER, H. REICHENBERGER and H. WENZEL: Z. Naturforschung $\underline{26a}$, 1452 (1971)

58. P. WISSMANN: unpublished.

59. H. REICHENBERGER: Dissertation Universität Erlangen-Nürnberg 1970

   a) see /57a/   b) see /57b/   c) see /36/

60. U. RICHTER: Dissertation TU Hannover 1964.

61. C. WÖLFING: Dissertation TU Hannover 1967.

62. H.G. KOCK: Dissertation TU Hannover 1967.

63. W. WIEBAUER: Dissertation Universität Erlangen-Nürnberg 1974

   a) G. WEDLER and W. WIEBAUER: Thin Solid Films, $\underline{28}$, 65 (1975).

64. J. FEDER, P. RUDOLF and P. WISSMANN: Proceedings of the 3rd Intern. Conf. on Thin Films, Budapest 1975.

65. H. PAPP: Dissertation Universität Erlangen-Nürnberg 1972

   a) G. WEDLER, G. FISCH and H. PAPP: Ber. Bunsenges. Phys. Chem. $\underline{74}$, 186 (1970)

   b) G. WEDLER and H. PAPP: Z. Phys. Chem. (Frankfurt) $\underline{82}$, 195 (1972)

   c) G. WEDLER, H. PAPP and G. SCHROLL: Surface Sci. $\underline{44}$, 463 (1974)

66. G. SCHROLL: Dissertation Universität Erlangen-Nürnberg 1972

   a) see /65c/

   b) G. WEDLER and G. SCHROLL: Z. Phys. Chem. (Frankfurt) $\underline{85}$, 216 (1973)

67. Y. MIZUSHIMA: Dissertation TU Hannover 1960

a) Y. MIZUSHIMA: J. Phys. Soc. Japan 15, 1614 (1960)

b) R. SUHRMANN, Y. MIZUSHIMA, A. HERRMANN and G. WEDLER: Z. Phys. Chem. (Frankfurt) 20, 332 (1959)

68. F.J. BRÖCKER: Dissertation TU Hannover 1967

a) F.J. BRÖCKER and G. WEDLER: Disc. Faraday Soc. 41, 87 (1966)

b) G. WEDLER and F.J. BRÖCKER: Z. Phys. Chem. (Frankfurt) 75, 299 (1971)

69. G. FISCH: Dissertation Universität Erlangen-Nürnberg 1970

a) see /65a/

b) G. WEDLER and G. FISCH: Ber. Bunsenges. Phys. Chem. 76, 1160 (1972)

70. P. WISSMANN and D. DAYAL: Thin Solid Films, to be published.

71. G. WEDLER and M. FOUAD: Z. Phys. Chem. (Frankfurt) 40, 1 (1964).

72. F.D. SNELL and C.T. SNELL: Colorimetric Methods of Analysis. New York: 1949, Vol. II.

73. R. SUHRMANN, R. GERDES and G. WEDLER: Z. Naturforschung 18a, 1208 (1963).

74. H.J. BUSSE: Dissertation TU Hannover 1962.

75. P. WISSMANN and H. GEIGER: to be published.

76. H. GENTSCH: Z. Phys. Chem. (Frankfurt) 24, 55 (1960).

77. P.A. REDHEAD, I.P. HOBSON and E.V. KORNELSEN: The Physical Basis of Ultrahigh Vacuum. London: Chapman and Hall Publ., 1968, p. 369 ff

78. P. LINDER: Dissertation Universität Erlangen-Nürnberg 1970.

79. R. GERDES: Dissertation TU Hannover 1963.

80. R.W. VOOK and F. WITT: J. Vac. Sci. Technol. 2, 49 (1965).

81. G.W. WILSON and B.P. SINHA: Thin Solid Films 8, 207 (1971).

82. K. MÜLLER and H. VIEFHAUS: Z. Naturforschung 21a, 1728 (1966).

83. A. KASHETOV and N.A. GORBATYI: Soviet Physics Solid State 10, 1673 (1969).

84. C.C. CHANG: Surface Sci. 23, 283 (1970).

85. R. NOSSEK: remarks in discussion, see reference 11, p. 559.

86. A.F. MAYADAS and M. SHATZKES: Phys. Rev. B 1, 1382 (1970).

87. K. BESOCKE and H. WAGNER: Phys. Rev. B (USA) 8, 597 (1973).

88. M.S.P. LUCAS: J. Appl. Phys. 36, 1632 (1965).

89. H. JURETSCHKE: J. Appl. Phys. 37, 435 (1966).

90. A.A. COTTEY: Thin Solid Films 1, 297 (1968).

91. R. ENGLMAN and E.H. SONDHEIMER: Proc. Phys. Soc. 69, 449 (1956).

92. P.J. PRICE: IBM J. Res. Develop 4, 152 (1960).

93. F.S. HAM and D.C. MATTIS: IBM J. Res. Develop 4, 143 (1960).

94. K.E. SAEGER and R. LÜCK: Phys. Kondens. Mat. 9, 91 (1969).

95. H.A. MÜSER: Phil. Mag. 45, 1237 (1954).

96. G. BRÄNDLI and P. COTTI: Helv. Phys. Acta 38, 801 (1965).

97. J.E. PARROT: Proc. Phys. Soc. 85, 1143 (1965).

98. R.M. MORE and D. LESSIE: Phys. Rev. B 8, 2527 (1973).

99. J.M. ZIMAN: Electrons and Phonons. Oxford: Clarendon Press, 1960, p. 459.

100. R.F. GREENE: in M. Green (ed): Solid State Surface Science. New York: Marcel Dekker Publ., 1969, p. 87 ff.

101. S.B. SOFFER: J. Appl. Phys. 38, 1710 (1967).

102. Y. NAMBA: Japan J. Appl. Phys. 9, 1326 (1970).

103. A. SEEGER: in S. Flügge (ed): Handbook of Physics. Berlin: Springer Verlag, 1955, Vol. 7, p. 427.

104. N.F. MOTT and J. JONES: The Theory of the Properties of Metals and Alloys. New York: Dover Publ., 1958, p. 289.

105. H. BROSS: in A. Seeger (ed): Moderne Probleme der Metallphysik. Berlin: Springer Verlag, 1965, Vol. 1, S. 400.

106. A. HAUG: Theoretische Festkörperphysik. Wien: Franz Deuticke Verlag, 1970, Band 2, p. 196.

107. R. NOSSEK: see reference 11, p. 550.

108. R.G. CHAMBERS: in I.M. Ziman (ed): The Physics of Metals, Part I: Electrons. Cambridge: University Press, 1969, p. 186.

109. Z. BASTL: Surface Sci. 30, 245 (1972).

110: W.A. HARRISON: Solid State Theory. New York: Mc Graw-Hill, 1970, p. 194 ff.

111. H. JONES: in S. Flügge (ed): Handbook of Physics. Berlin: Springer Verlag, 1956, Vol. 19, p. 266 ff.

112. A.N. GERRITSEN: see reference 111, p. 206 ff.

113. J.M. ZIMAN: see reference 108, p. 275 ff.

114. D.K.C. MAC DONALD: in S. Flügge (ed): Handbook of Physics. Berlin: Springer Verlag, 1956, Vol. 14, p. 137 ff.

115. R. LÜCK: Z. Metallk. 61, 965 (1970).

116. J.S. DUGDALE and L.D. FIRTH: Phys. Kondens. Mat. 9, 54 (1969).

117. J. FRIEDEL: see reference 108, p. 389.

118. LANDOLT-BÖRNSTEIN: Zahlenwerte und Funktionen. Berlin: Springer Verlag, 1964, Band IV,Teil 2b, S. 327.

119. R.W. SPRINGER and R.W. HOFFMAN: J. Vac. Sci. Technol. 10, 238 (1973).

120. K. EYRICH: Diplomarbeit Universität Erlangen-Nürnberg 1973.

121. J.M. ZIMAN: see reference 99, p. 113

122. V.A. MARSOCCI and S.S. SHUE: J. Appl. Phys. 42, 5047 (1971).

123. H.E. BENNETT, J.M. BENNETT, E.J. ASHLEY and R.J. MOTYKA: Phys. Rev. 165, 755 (1968).

124. A. LAL and V. DUGGAL: Thin Solid Films 14, 349 (1972).

125. R. NOSSEK: Z. Naturforschung 16a, 1162 (1961).

126. E.E. MOLA, I. BORRAJO and J.M. HERAS: Surface Sci. 34, 561 (1973).

127. R.E. HUMMEL: Optische Eigenschaften von Metallen und Legierungen. Berlin: Springer Verlag, 1971, p. 194.

128. S. RITZ: Dissertation Universität Erlangen-Nürnberg 1975.

129. H. SUGAWARA, T. NAGANO, K. UOZUMI and A. KINBARA: Thin Solid Films 14, 349 (1972).

130. D.B. TANNER and D.C. LARSON: Phys. Rev. 166, 652 (1968).

131. J.T. JACOBS, R.C. BIRTCHER and R.N. PEACOCK: J. Vac. Sci. Technol. 7, 339 (1970).

132. A. BERMAN and H.I. JURETSCHKE: Appl. Phys. Lett. 18, 415 (1971).

133. R. GRIGOROVICI and G. CIOBANU: see reference 11, p. 596.

134. H. BERGER, W. KAHLE and G. JAENICHE: in E. Hahn (ed): Proceedings of the II. Colloquium on Thin Films, Budapest 1968, p. 408.

135. J.G. SWANSON, D.S. CAMPBELL and J.G. ANDERSEN: Thin Solid Films 1, 325 (1967/68).

136. G. WEDLER and P. WISSMANN: Thin Solid Films 2, 391 (1968).

137. Z.H. MEIKSIN and R.A. HUDZINSKI: J. Appl. Phys. 38, 4490 (1967).

138. M.C. MARTIN: J. Appl. Phys. 41, 5163 (1970).

139. A. SINGH: Thin Films 2, 159 (1972).

140. B.S. VERMA and S.K. SHARMA: Thin Solid Films 5, R 33 (1970).

141. E. KLOKHOLM: J. Vac. Sci. Technol. 10, 235 (1973).

142. J.L. OLSEN: Helv. Phys. Acta 31, 713 (1958).

143. F.J. BLATT and H.G. SATZ: Helv. Phys. Acta 33, 1007 (1960).

144. M.Ya. AZBEL and R.N. GURZHI: Sov. Phys. JETP 15, 1133 (1962).

145. J.B. van ZYTVELD and J. BASS: Phys. Rev. 177, 1072 (1969).

146. J.S. DUGDALE and Z.S. BASINSKI: Phys. Rev. 157, 552 (1967).

147. A. KINBARA and K. UEKI: Thin Solid Films 12, 63 (1972).

148. T.S. JAYADEVAIAH: Thin Solid Films 4, R 37 (1969).

149. J. le BAS: Thin Solid Films 10, 437 (1972).

150. F.J. BLATT: Proc. Phys. Soc. London A 83, 1065 (1964).

151. LANDOLT-BÖRNSTEIN: Zahlenwerte und Funktionen. Berlin: Springer Verlag, 1959, Band II, Teil 6, p. 939.

152. A. SOMMERFELD and H. BETHE: Elektronentheorie der Metalle. Berlin: Springer Verlag, 1967, p. 191.

153. J. GOUAULT: J. Physique (France) 28, 931 (1967).

154. N.I. IONESCU: Z. Phys. Chem. (Frankfurt) 78, 108 (1972).

155. J. HORIUTI and T. TOYA: in M. Green (ed): Solid State Surface Science. New York: Marcel Dekker, 1969, p. 1 ff.

156. W. KIRSTEIN: Dissertation Universität Hamburg 1974.

157. M. WATANABE: Surface Sci. 34, 759 (1972).

158. R.F. GREENE and R.W. O'DONELL: Phys. Rev. 147, 147 (1966).

159. T. SUGITA, S. EBISAWA and K. KAWASAKI: Surface Sci. 11, 159 (1968).

160. O.A. PANCHENKO, P.P. LUTSISHIN, Yu.G. PTUSHINSKIJ and V.V. SHISHKOV: Surface Sci. 34, 187 (1973).

161. E. KRAMP: Dissertation TU Hannover 1966.

162. R. SUHRMANN, M. KRUEL and G. WEDLER: Z. Naturforschung 18a, 633 (1963).

163. C. REALE: Phys. Lett. A 43, 239 (1973).

164. J.R. CHAUVINEAU and C. PARISET: Vide (France) 28, 60 (1973).

165. T.C. BOYCE and W.H. WONG: Phys. Lett. 36A, 323 (1971).

166. K.P. GEUSS: Dissertation Universität Erlangen-Nürnberg 1974.

167. G. EHRLICH: J. Chem. Phys. 35, 2165 (1961).

168. P. LÖCHNER: Diplomarbeit Universität Erlangen-Nürnberg 1974.

169. T. BAUER and P. WISSMANN: Ber. Bunsenges. Phys. Chem. 77, 1024 (1973).

170. J.C. TRACY: J. Chem. Phys. 56, 2736 (1971).

171. H. MADDEN, J. KÜPPERS and G. ERTL: J. Chem. Phys. 58, 3401 (1973).

172. K. CHRISTMANN, O. SCHOBER and G. ERTL: J. Chem. Phys. 60, 4719 (1974).

173. S. LIPPMANN: in: Elektronenstruktur und physikalische Eigenschaften metallischer Werkstoffe. Leipzig: VEB Deutscher Verlag für Grundstoffindustrie, 1972, p. 255.

174. H.J. BAUER: Z. Phys. 177, 1 (1964).

175. B. BARANOWSKI, K. BOCHENSKA and S. MAJCHRZAK: Roczniki Chemii, Ann. Soc. Chim. Polonorum 41, 2071 (1967).

176. J.M. ZIMAN: see reference 99, p. 337.

177. R. WISNIEWSKI: Phys. Stat. Sol. 5, pK 31 (1971).

178. B.R. COLES: Proc. Phys. Soc. B 65, 221 (1951).

179. Z. BASTL: Surface Sci. 22, 465 (1970).

180. R. SUHRMANN, G. SCHUMICKI and G. WEDLER: Z. Naturforschung 19a, 1208 (1964).

181. G. WEDLER and G. SANTELMANN: Ber. Bunsenges. Phys. Chem. 75, 1026 (1971).

182. O. BEECK, W.A. COLE and A. WHEELER: Disc. Faraday Soc. 8, 314 (1950).

183. M. WAHBA and C. KEMBALL: Trans. Faraday Soc. 49, 1351 (1953).

184. D.F. KLEMPERER and F.S. STONE: Proc. Roy. Soc. (London) A 243, 375 (1957).

185. E. RIDEAL and F. SWEETT: Proc. Roy. Soc. (London) A 257, 291 (1960).

186. D. BRENNAN and F.H. HAYES: Trans. Faraday Soc. 60, 589 (1964).

187. K. CHRISTMANN, O. SCHOBER, G. ERTL and M. NEUMANN: J. Chem. Phys. 60, 4528 (1974).

188. J. BAGG and F.C. TOMPKINS: Trans. Faraday Soc. 51, 1071 (1955)

189. H. STROTHENK: Diplomarbeit TU Hannover 1961.

190. G.C. BOND: Catalysis by Metals. London: Academic Press, 1962, p. 7 ff.

191. D.C. LARSON and B.T. BOIKO: Appl. Phys. Lett. 5, 1382 (1964).

192. K.L. CHOPRA: Phys. Rev. 155, 660 (1967).

193. J.F. KOCH: Phys. Kondens. Mat. 9, 146 (1969).

194. Yu.G. GAIDUKOV and I. KADLECOVA: Phys. Kondens. Mat. 9, 192 (1969).

195. H. SCHWARZ and R. LÜCK: Mater. Sci. Eng. 5, 149 (1969/70).

K. Müller

How Much Can Auger Electrons Tell Us About Solid Surfaces?

1. Introduction

Science and technology have developed a still increasing interest
in surfaces of solids. The information which is wanted about the few
outermost atom layers, the selvedge, includes the arrangement of
atoms, chemical composition as well as electronic properties. Among
the many different interactions between particles and solids which
can be successfully studied for this purpose /1/ only those shall
be considered in this paper which are accompanied by the emission of
slow electrons of characteristic energy from the surface. Low energy
electrons (10....1000 eV) travel a mean free path of the order of
10 Å between two events of interaction with the solid. Therefore they
can be emitted without further inelastic collisions only if origina-
ting from the selvedge. That is what gives slow electrons their surface sensitivity and hence their importance in surface physics.

This paper on Auger electron spectroscopy (AES) was written as
a brief outline of the general aspects and implications of AES with
a side-glance to related methods. It is an attempt to draw attention
on some potential future applications of AES rather than a detailed
review of the success that has already been achieved /2,3/. Conse-
quently the selection of references is somewhat arbitrary and no
attempt for completeness has been made.

Following a discussion of various processes which lead to discrete lines in the spectrum of emitted electrons, and after some remarks
on instrumentation a spectrum of Auger electrons is presented for
inspection of its characteristic features. Peak position, peak intensity as well as the shape of the features in the spectrum will be
discussed with respect to qualitative element analysis, quantitative
aspects and electronic structure of the specimen. A possibility for
the identification of simple compounds will be proposed.

2. The Distribution N (E)

First of all the surface must be activated to emit slow electrons. This can be done by electron bombardment with an energy E_p of the order of 1 keV or by the interaction with soft X-rays. In case of electron impact the secondary electron distribution $N(E)$, the total number of electrons backscattered into the hemisphere and the energy interval E... E + dE, consists of three characteristic parts: At the high energy end of the spectrum dominates a narrow peak of elastically and quasi-elastically scattered electrons. These are used extensively by low energy electron diffraction (LEED) for the determination of surface structure. The low energy side consists of the so called "true" secondaries which are emitted at the end of a sequence of scattering processes in which they have suffered multiple energy losses. This is a broad maximum as compared to the elastic peak. Both parts join through a more or less flat background to a continuous distribution. A spectrum of small discrete lines is superimposed which contains two types of peaks: Those which move on changing primary energy along the energy scale by the same amount E_p is altered, and those whose position does not depend on E_p. It is this discrete spectrum that must be directly correlated to the electron states in the specimen.

3. Energy Levels of the Sample

Imagine a highly simplified scheme of electron energy levels of a solid as it is drawn in the center of fig. 1. On a vertical arbitrary scale the scheme contains several discrete levels (E_1, E_2), the core states of the atoms, far below the Fermi level E_F, all of which are occupied by electrons. There is a valence band represented by energy E_3. Another band of unoccupied bound states above but close to the Fermi level is provided and represented by E_4. The scheme is laterally divided into five identical sections which are used to explain some of the many possible transitions between bound states and the continuum of free states above the vacuum level. Each one of them is fundamental for a method useful in surface physics. The primary electrons of energy E_p = eU are supplied by the emitter, the back scattered electrons are collected by the detector. Both of them are represented by their inherent energy levels on the left and right hand side of fig. 1, respectively. The scheme does not restrict our discussion to targets that are insulators or semiconductors as drawn. It is easy to imagine further states forming a conduction band about E_F.

4. Some Important Interactions

4.1 Inelastic Loss Spectroscopy (ILS)

It begins with the excitation of bound electrons into unoccupied bound or free states by interaction with the primary beam. Since an incident electron above a threshold energy can in principle exchange any amount of its kinetic energy, the electrons which are set free from the solid are of no particular interest. They contribute to the background of the secondary electron distribution $N(E)$. If, however, an electron from one of the core states or occupied bands is lifted into an empty bound state close to the Fermi level, then the corresponding primary electron has suffered a discrete energy loss ΔE /4, 5,6/. The search for inelastic loss peaks in $N(E)$ is called inelastic loss spectroscopy (ILS) which is schematically shown in the first section of fig. 1. Since in solid state physics all energy levels are given with respect to the Fermi level and empty states are close to it, ΔE has the approximate value of the electron binding energy for a certain state. In other words: While most electrons of the primary beam ionize the atoms of the target in various core levels without giving direct evidence for this process in the spectrum, some of the primary electrons do indeed indicate the levels where they created a hole by an inelastic loss peak. This is quite valuable information which we will call upon later in the context of Auger electrons. It is clear that inelastic losses generally occur by transfer of energy to all quasiparticles such as phonons and especially by excitation of plasmons /7,8/. The inelastic losses represent that part of the discrete spectrum that moves together with the primary energy.

4.2 Disappearance Potential Spectroscopy (DAPS)

The following two parts of fig. 1 display the situation where the primary energy E_p = eU just reaches the threshold to elevate core electrons to unoccupied states E_4 near E_F, in the second section they originate from the level E_2, in the third section from the deeper lying core level E_1. In such a case the primary electrons involved populate free states near E_4 and thus vanish from the spectrum /9/. Disappearance potential spectroscopy (DAPS) indicates thresholds for

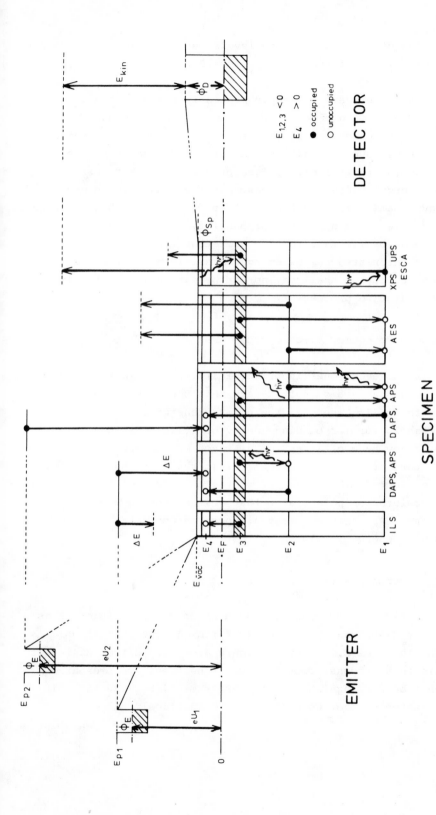

Fig.1 Simplified scheme of energy levels of a solid in interaction with free electrons and photons. For abbreviations see text.

the ionization of consecutive core levels as a function of the acceleration potential U of the primary electrons.

In any case there are now empty states in certain core levels of the target. The excited electronic system of the solid must return to the ground state by electron transitions from occupied levels to the empty states. Such transitions can either be accompanied by radiation or appear to be radiationless processes which transfer their transition energy to other electrons. For low transition energies, E < 1000 eV (that is for light elements or transitions between higher levels of heavier elements), the probability for the radiationless process exceeds by far that for the competing radiative process /10/. Since we are interested in low energy transitions in order to gain surface sensitivity we shall pay no further attention to the less favourable radiative process.

4.3 Appearance Potential Spectroscopy (APS)

There is, however, one (in principle!) very simple technique based on the appearance of soft X-rays that deserves our attention in this context /11/. Whenever the primary energy exceeds the ionization threshold for (or binding energy of) electrons in a certain core level a new beam of X-rays of distinct energy appears. This radiation in turn produces photoelectrons at the surrounding surfaces which are collected. The number of photoelectrons increases stepwise at every acceleration potential U that corresponds to a discrete binding energy. Appearance potential spectroscopy (APS) opens one way of determining energy levels in solids.

4.4 Auger Electron Spectroscopy (AES)

The radiationless process already mentioned is the Auger transition which is displayed in section 4 of fig. 1. The initial hole is taken to be in level E_1. Generally there are three different levels involved in this process. Two distinct transitions are presented which cannot be distinguished by the experiment, because the initial and final states of the two processes are identical.

4.5 Photoelectron Spectroscopy (XPS, UPS)

The last section of fig. 1 shows what happens when a photon as a carrier of the primary energy is absorbed. Provided the photon energy is known, the emitted photoelectrons give immediate information on the energy levels of the solid. In case of low energy photons (UV light) mainly band states are probed. X-rays are used for investigation of core states. Electron spectroscopy for chemical analysis (ESCA) /12/ (or X-ray respectively ultraviolet photoelectron spectroscopy, XPS, UPS) is a complementary method to the spectroscopy of Auger electrons (AES).

5. Instrumentation

It has become apparent from the discussion above, that the detector on the right hand side of fig. 1 must actually be a spectrometer. Several different electrostatic instruments are in use /2/. We shall briefly introduce two examples whose basic conceptions compare as follows. The retarding field analyzer (RFA) collects all electrons with sufficient energy to surmount a variable retarding field E_r. Thus its output current $I_{RFA}(E_r)$ is proportional to the area under the distribution curve $N(E)$ beginning at E_r

$$I_{RFA}(E_r) = C_{RFA} \cdot \int_{E_r}^{E_p} N(E) \ dE$$

where E is proportional to E_D. The spectrum itself is conveniently measured by electronic differentiation.
Dispersive types of electrostatic analyzers deflect the electrons according to their energy. They differ from each other in the symmetry of the variable deflection field E_D and their focusing properties.

Widely used is the cylindrical mirror analyzer (CMA), whose output current is directly proportional to the wanted energy distribution N(E) times the energy E

$$I_{CMA}(E_D) = C_{CMA} \cdot E \cdot N(E)$$

In order to detect the faint discrete structures superimposed on the strong background of N(E) both types of instruments are operated in a mode as to record an output current proportional to dN(E)/dE. In case of the RFA this is achieved by the second derivative of I_{RFA}. The CMA, however, leaves us after differentiation of its output with the modified spectrum

$$dI_{CMA}/dE \equiv (dN(E)/dE)^* = C_{CMA} \cdot (E \cdot dN(E)/dE + N(E))$$

This distribution does qualitatively exhibit the same structure as does dN(E)/dE itself. It is enhanced at higher energies with respect to dN(E)/dE because of the factor E and shows a higher background at low energies because of the additive N(E). The spectrum $(dN(E)/dE)^*$ is directly plotted by the CMA and is presented in most papers (and also here) without the asterisk.

Both types of instruments have been developed to a comparable resolution of better than 1 %. Because of the fact that the RFA originally collects all electrons above a variable threshold while the CMA only detects those within a narrow interval about a variable energy the latter instrument has a much better signal/noise-ratio, almost by a factor of 100. Equivalent to this is a much faster scan rate for the CMA which makes it possible to display the spectrum on an oszilloscope.

6. Auger Transitions

Let us now take a closer look at Auger electrons. They have first been recognized as discrete features in energy spectra of secondary electrons by LANDER /13/. As has already been explained, the Auger deexcitation of the electron system does not involve primary electrons and thus is independent of E_p. The initial vacancy in a core state is filled by an electron transition from a higher level while the energy set free by this transition is transferred to another electron. If the transition energy exceeds its binding energy, the Auger electron can be released from the surface. It can escape without inelastic collisions, and then carries away a characteristic kinetic energy E_{kin}. According to fig. 1, the overall energy balance is

$$E_2 - E_1 = E_F - E_3 + E_{kin} + \Phi_D \qquad (*)$$

The usual nomenclature is adopted from the physics of atoms, where the energy levels are designated by electron shells K, L, M ... . For an Auger transition, the shell with the original hole is stated first followed by the shells of the initial states of the electrons involved. For example, KL_1L_2 names an Auger transition with an initial hole in the K shell to be filled by an L_1-subshell electron and a L_2-electron to be emitted. As has been pointed out, the experiment does not distinguish between KL_1L_2 and KL_2L_1 transitions since the initial and final electron configurations do not differ in both cases. If valence band electrons of a solid are involved, V stands for a valence band level.

The Auger deexcitation is a complex process that is by far not yet fully understood. Even for free atoms the determination of transition probabilities is generally too complicated to be solved since electron-electron interaction of the two electrons involved plays the important role in the process. Only for heavy atoms where the assumption of jj-coupling is adequate and for high transition energies, it has been possible to predict transition rates /14/, and some of the results were experimentally conformed. In the context of surface physics we are most interested in the low energy cases for light elements and individual outer shell transitions which, however, have not yet been theoretically treated. There are particularly no

explicit calculations of Auger transition probabilities in solids
available where the $\vec{k}$-dependency of the matrix elements involving
band states and many-particle effects make the problem even more
complicated.

7. Inspection of an Auger Spectrum

Despite of the desperate theoretical situation one observes clear
Auger electron spectra which exhibit distinct features. Fig. 2 shows
part of a derivative spectrum dN(E)/dE with its typical double wing

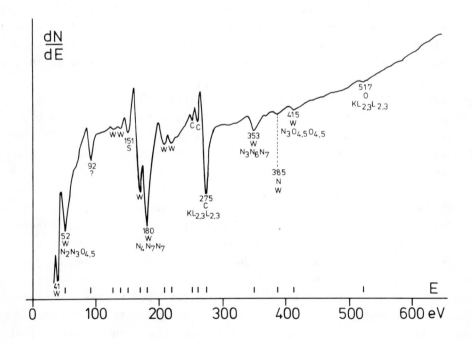

Fig.2   Auger spectrum of a tungsten sample with a few contaminants. For the
        ?-peak see Fig.9.

peak shapes taken from a tungsten sample. The information offered by
the experiment is concealed in

a) peak positions
b) peak intensities and
c) shape of the features in the spectrum.

Because of the inherent width of each peak its <u>position</u> is determined
(by convention) at the minimum of its lower wing. Only the peak positions are used in qualitative element analysis. The <u>intensity</u> must
be a quantitative measure for the number of atoms involved in a particular transition and hence is of urgent interest in quantitative
analysis. Finally the <u>peak shape</u> comprising a possible fine structure
of the feature indicates nonresolved transitions involving split core
levels or reflects the density of states in a band. Also influences
of the chemical environment of the atoms are sometimes seen in peak
shapes.

Qualitative analysis, steps towards quantitative analysis and the
line shape information will now be discussed separately.

8. The Qualitative Element Analysis

If the electron energy levels in solids were fixed values known
for all materials, the qualitative analysis of a specimen would be
easy routine. In principle it just needs the determination of Auger
peaks in the spectrum and their comparison with tabulated values of
binding energies by virtue of equation (*). Unfortunately, this data
collection is not directly available, and well known X-ray levels of
atoms can only serve as a first approximation. Their application is
not really adequate since photons leave their source atom in a different final state than Auger electrons do. As a consequence of Auger
emission the electron density and the shielding effect of the electron
cloud on the ion core change. Therefore at least the value for the

initial state energy E_i of the Auger electron must be altered. A proposal by CHUNG and JENKINS /15/ introduced a satisfactory correction which replaces $E_i(Z)$ by the average of corresponding X-ray levels for elements of atomic number Z and Z + 1,

$$E_i'(Z) = (E_i(Z) + E_i(Z + 1))/2.$$

Since generally two of the three levels involved in an Auger transition can serve as initial state of the electron to be emitted (levels E_2 and E_3 in fig. 1, section 4), both of them must be treated in the given manner. The corrected values have been filed /16/ and serve as basic data to predict the position of possible Auger peaks. The agreement with experimentally observed transitions is surprisingly good, deviations are of the order of a few eV. In cases of doubt it is sometimes helpful to determine the primary energy threshold for the peak under consideration in order to find the initial hole state for the transition. The complementary use of ILS is adventageous in this context. Most Auger features can thus be properly identified and be associated with the presence of certain elements on/in the sample surface. It is this property of a fingerprint that gives Auger spectra their importance.

As a typical example fig. 2 shows part of a tungsten spectrum which was taken with a CMA using 1000 eV primary energy. Various outer shell transitions of W are well resolved, some of which are explicitly labelled. Inner shell transitions beginning with the M series (MNN ...) could not be excited by the primary energy. Further inspection of the curve indicates several quite common contaminants: Carbon shows up strongly with its KLL series, sulphur is present as well as traces of oxygen. The identification of nitrogen is doubtful in this case because its KLL-peak at 385 eV coincides with a possible tungsten transition. A slight shoulder at 67 eV and the strong peak at 92 eV cannot be easily identified. Although the latter transition energy usually indicates the presence of silicon, other Si peaks fail to show up. It is probably fair to say that in most cases the constituents of a surface can be decidedly recognized. The analysis so far does indeed involve only the peak positions along the energy scale in the spectrum.

Sometimes, however, one is left with uncertainties because of the coincidence of Auger energies resulting from transitions of dis-

tinct elements. Difficulties arise when peaks appear at energies
several eV apart from the expected values. This of course can happen
since the energy level correction did not take the solid into account
to which the emitting atom is bound. Occasionally there are even
peaks which remain totally unidentified under the simple rules of
element analysis. However, these and other features in the Auger
spectra may open new applications of AES beyond the element analysis.

9. Steps Towards Quantitative Analysis

The problem of quantitative analysis raises two questions:
(i)   How can the total amount of "true" Auger electrons (resulting
      from a certain element) be retrieved from the background?
(ii) How does the number of Auger electrons collected by the detec-
      tor relate to absolute atomic densities in the selvedge of the
      specimen?

First of all the distribution N(E) must be generated by integration, if it is not directly recorded. Several attempts of background
determination and subtraction have been made ranging from simple
guesses to calculations using a polynomial fit /17,18,19/. A problem
results from the question how far the generally asymmetric Auger peak
"extends" to the low energy side /20/. Not only the Auger electrons
directly resulting from a transition must be taken into account but
also those which have suffered subsequent losses. There is some arbi-
trariness involved at this point but methods are improving.

Fig. 3 is qualitatively taken from the work of STAIB and KIRSCHNER
/19/. It shows a mica spectrum containing peaks of silicon, sulphur
potassium and oxygen. The derivative spectrum dN(E)/dE as the output
of the spectrometer was integrated and the background under the peaks
(indicated by a dotted line) was determined by means of a spline

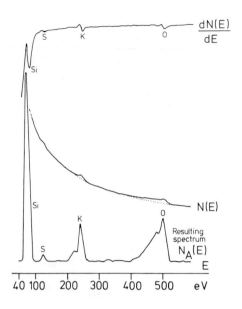

Fig.3 Quantitative Auger analysis of
a mica surface (drawn from /19/).
dN(E)/dE as recorded, N(E) after
integration. Calculated background indicated by dotted line.
$N_A(E)$ after background subtraction.

approximation. Subsequent background subtraction results in the distribution $N_A(E)$ of true Auger electrons. Finally the area under each
peak consisting of multiplets and loss features is a measure of the
Auger current which results from the corresponding transitions.

This Auger current I_A is proportional to several parameters:

$$I_{XYZ} = K \, \Omega \, I_p \, (1 + r) \, a \, \sigma_X(E_p) \, T_{XYZ} \cdot n$$

K is an experimental constant, Ω measures the angle of acceptance of
the spectrometer, I_p is the primary current and r its back scattering
factor, a resembles the attenuation by scattering processes after the
transition. $\sigma_X(E_p)$ is the ionization cross section for electron induced ionization of the X level, T_{XYZ} the transition probability for
the whole X series of transitions and finally n stands for the desired
atomic density. While $\sigma(E_p)$ can be measured (see fig. 4 as an example

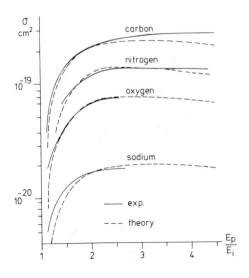

Fig.4 Cross sections for electron
impact ionization of the
K-shell (after /21/).

/21/), T_{XYZ} is assumed to be unity because of the extremely low photon emission probability in the low energy range. The geometrical
factors and attenuations can be determined experimentally and have
been thoroughly considered /21,22/. As a result it was possible to
calculate from the distribution $N_A(E)$ in fig. 3 the absolute density
of K atoms in a cleavage plane of mica to within 20 % of the known
value. To indicate the sensitivity of the method the Auger current
in fig. 3 of the order of 10^{-10} A was generated by K atoms of approximate density of $2\cdot10^{14}$ atoms/cm^2 which were excited by a primary
current of $7\cdot10^{-6}$ A. Generally one assumes that the signal from a
fraction of 10^{-2} of a monolayer can be detected.

Even without going through the details discussed above, relative
measurements can be made. It has been shown for various simple specimen that peak to peak amplitudes of dN(E)/dE as well as peak hights
in the N(E) distribution can be calibrated quite accurately by comparison with a quartz crystal oscillator to give quantitative information on atom density and layer thickness of a condensate /23/.

10. Deconvolution

Before turning to the shape of Auger peaks care must be taken to separate instrumental broadening and inelastic losses from the true Auger electron distribution $N_A(E)$ by proper deconvolution methods /24/. The experimental curve $N(E)$ recorded by the spectrometer is given by the convolution integral

$$N(E) = \int\limits_{-\infty}^{+\infty} N_A(E') \cdot N_i(E-E') dE',$$

the measurement, of course, is restricted to an appropriate finite energy range.

The broadening function $N_i(E)$ which contains finite instrument resolution and inelastic losses must be known in order to retrieve $N_A(E)$ from the spectrum. Assuming that a monoenergetic beam of primary electrons suffers the same broadening as the Auger lines do, one finds $N_i(E)$ as the elastic peak and its low energy tail taken at an energy within or close to the interval of interest. Fig. 5 shows part of a platinum Auger spectrum $N(E)$ recorded by a retarding field analyzer (curve a). The measured $N_i(E)$ is displayed by b) and c) is the result of deconvolution $N_A(E)$. It definitely displays more prominent details of the spectrum which were obstructed by instrumental broadening and inelastic losses.

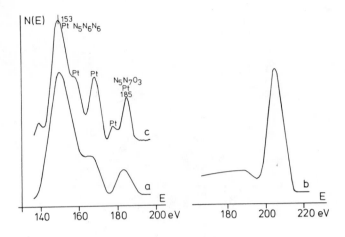

Fig.5   The effect of deconvolution: a) part of a platinum
        spectrum N(E) as recorded, b) elastic peak containing
        broadening effects, c) resulting spectrum after de-
        convolution.

11. Line Shape and the Density of States

Let us assume a broad Auger peak of characteristic shape, which
has been treated by proper deconvolution. It still contains the inherent width of the energy levels due to lifetime broadening. Experimentally observed fine structure could result from core level split-
ting or inelastic loss processes. If this is definitely not so, the
peak probably consists of core level-band-transitions XVV or XYV. In
such a case the prominent peaks in the density of states can be re-
garded as quasi-discrete levels that can be involved in individual
Auger transitions of the types indicated in fig. 6.

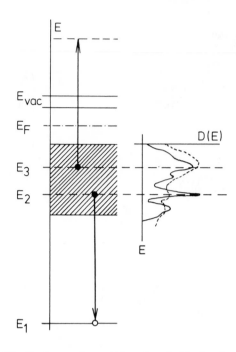

Fig.6 Auger transition involving band
states with corresponding density
of states D(E), experimental result (dashed curve), theoretical
distribution (solid curve).

The number of electrons emitted as a result of a transition XVV
is given by the matrix element square for this process times the
"combined" density of states $D^*(E)$ which is a convolution integral of
the density-of-states-function at different levels within the band
/13/,

$$D^*(E) = \int_\Delta D(E+\Delta) \cdot D(E-\Delta) d\Delta$$

The energy parameter Δ varies within the width of the band. As has
already been pointed out in this paper very little information is
available on Auger transition probabilities with band levels involved.
Nevertheless it can be tried to recover the density of states from
the electron distribution $N(E)$. The procedure needs the assumption
of a constant matrix element for all transitions in which the one
core level X and various band levels take part. This assumption may
not always be a good approximation. If applicable, however, $N(E)$ is
essentially equal to $D^*(E)$ which can then be inverted to retrieve the
density of states $D(E)$.

114

Several attempts of this kind have been made especially for the valence band of silicon using LVV transitions /25,26/. A typical result is given on the right hand side of fig. 6 (dashed line) and compared with a theoretical distribution D(E), (solid line).

Whenever such a procedure is successful, Auger spectroscopy offers electronic data of surfaces which are otherwise difficult to obtain /27,28,29/. Of course UPS and also ion neutralisation spectroscopy (INS) /30/, are complementary methods to apply for comparison of results, but they need a higher degree of sophistication of the instruments than AES normally does.

Recalling the different methods and interactions compared in fig. 1 it is obvious that among those only AES, UPS and XPS can explore the density of occupied states. ILS, DAPS and APS, however, all need a free bound state in the specimen and thus probe the density of empty states.

12. Line Shape and Chemical Environment

In surface physics even more interesting than the density of states itself is its variation with surface conditions. Already the experiments with silicon indicated that the cleanliness of the surface strongly affects D(E) /26/. The number as well as the energy of the transitions belonging to the Auger peak vary with the state of oxydation and other kinds of surface reactions. So far Auger electron spectroscopy had proven its applicability as a method for element analysis, in which a peak in the spectrum is associated with a certain kind of atom. Now it turns out that at least the fine structure of Auger peaks is sensitive to chemical reactions /31/. A detailed study of the fine structure in the Auger spectrum of sulphur and oxygen on a nickel surface similar to the oxygen-silicon experiments

showed again the variation of peak shapes with chemisorption /32/.
The changes arise from transitions involving surface molecular orbitals formed by electrons in chemisorption bonds between the adsorbed
species and the outermost surface atoms. The localized density of
states $D_\ell(E)$ at adsorption sites will thus be different from the
"clean" and certainly different from the bulk density of states.

The above qualitative interpretation also holds in cases where
the outer shell electrons are not necessarily united in a band. Carbon, for example, displays its KLL Auger series with different fine
structur according to the chemical bonds to which the carbon L-shell
electrons contribute. Fig. 7 shows the same KLL peak of carbon present as a contaminant on a tungsten surface in two shapes. Curve a)

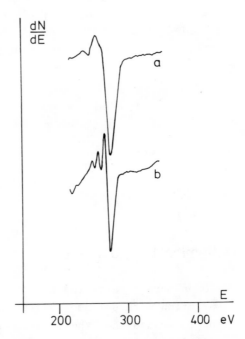

Fig.7   The carbon KLL Auger series in
different chemical environments:
a) C segregated on a W-surface,
b) C in a C-Cs compound (Details
from Fig. 9).

represents the case of carbon segregated at the surface. The lower
part b) shows the peak after a chemical reaction with cesium which
will be discussed in more detail in the next chapter. Many observations of this kind have been made which could be explained in terms
of bound states due to molecular orbitals /33/.

116

Also inner shell levels are affected by a chemical bond. Changes
of the order of a few eV known as chemical shifts are often observed
in XPS. It is difficult in AES to relate peak shifts to changes of
certain levels since three of them are involved /34/. Chemical shifts
are, however, at least indicative for chemical bonds also in AES.

13. Auger Electrons from Compound Solids

Auger spectra from compounds sometimes contain peaks which appear
to have shifted about (or more than) 10 eV with respect to the transition of one of the constituents. Some examples of an increasing list
are the oxides of Al /35/, Be /36/, Mg /37/, Li /38/ and Si /39/.
In other cases prominent peaks in the spectrum cannot be attributed
to either one of the components at all. A Cs-C compound will be presented as an example. Provided the peaks in question do not indicate
a contaminant, they are transitions characteristic of the compound.
Of course there is generally a possible explanation in terms of the
preceding chapter if one invokes unusually big changes in the density
of states or enormous chemical shifts. While this is not generally
accepted a slightly different interpretation in terms of cross transitions is conceivable.

14. The Concept of Cross Transitions

Suppose two distinct elements A and B are present at a surface.
There are two different schemes of energy levels and hence a spectrum
will be observed that may contain all possible transitions of each
of the components. Suppose further these components form a compound,
then in first approximation, the combined energy level scheme consists of the total of all levels of A and B, with their energies modified by normal chemical shifts. Because of the existence of binding
states an overlap of some of the wave functions ψ_A of eigenstates
E_A with wave functions ψ_B is assumed. It is conceivable then to expect
cross transitions between outer shell levels of the kind proposed
in fig. 8. On the left hand side there are two sets of energy levels

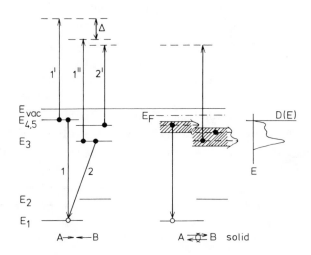

Fig.8 The concept of cross transitions. Left: A and B are energy level schemes
for different atoms. (1,1') regular Auger process. (1,1'') and (2,2') cross
transitions. Right: Cross transition involving band states in a solid.

A and B with an initial vacancy in the atomic core level E_1. A regular
Auger transition (1, 1') in atom A is drawn. Assuming interactions
between electrons populating levels of A and B the second stage of
transition 1 may also be 1" thus leading to a cross transition (1, 1").
The expelled electron would then appear to have "shifted" in energy

by Δ with respect to the transition (1, 1'). A second cross transition (2, 2') initiated by the same core level excitation emits an electron of similar energy.

The concept of cross transitions is conceivable also if band states of a solid are involved. Provided the local density of states varies from the environment of site A to that of B, a transition is proposed on the right hand side of fig. 8. In such a case distinct peaks of the total density of states D(E) are supposed to consist of electrons whose wave functions are of different symmetry. One peak may be populated by, say, s-electrons of A while in another peak d-electrons of B prevail.

15. A Potential Example: Cs-C

A curious peak in the spectrum of a tungsten surface has given rise to the assumption that it may be an example for a cross transition /40/. The spectrum of fig. 2 which is identical with curve c) in fig. 9 did already display this peak at 92 eV. Curve a) represents the W sample during cleaning procedure containing strong signals due to carbon and oxygen. Adsorption of cesium resulted in spectrum b) which shows several changes with respect to a). Low and high energy Cs lines appear as expected, the onset of a peak at about 90 eV and a definite change in the fine structure of the carbon peak are observed. Following a heat treatment up to 1000 K the Cs peak diminishes but the ?-peak becomes considerably stronger. The C peak fine structure remains, (curve c).

The general behaviour of the curious peak can be summarized by the following statements:

(i)   Both of the components C and Cs must be present for the peak to appear. Neither a C-contaminated sample that has not been

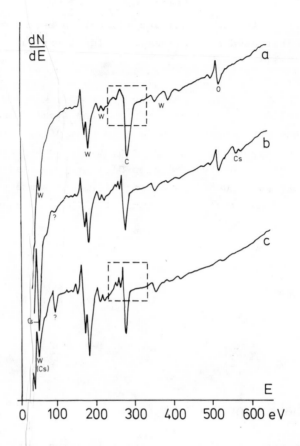

Fig.9 Auger spectra of a tungsten sample: a) during a cleaning procedure, contaminants carbon, sulphur, oxygen, b) with cesium condendate, c) after heat
treatment to 1000 K. Same spectrum as displayed in Fig.2. The 92 eV peak is
probably due to a cross transition between Cs and C. Details of the carbon
peak see Fig.7.

exposed to Cs nor the carbon free but Cs covered surface pro-
duce this peak.

(ii)   Increasing Temperature does increase peak intensity, it van-
ishes, however, at temperatures above 1400 K.

(iii)   The tungsten substrate does not appear to participate in the
electronic process.

There are compounds of C and Cs known which have the notation
$C_{4m}Cs$, m = 2, 3, 4 ... They have been observed up to $C_{60}Cs$.

In these compounds C is present in the graphite structure with the Cs atoms either in layers between the graphite layers or in dislocations /41/. Elevated temperature is believed to enhance the diffusion of Cs into the graphite layer structure. Compounds of high Cs-content transform to those of lower content during temperature activated Cs desorption. The most stable compounds of high carbon content seem to desorb above 1600 K having a strong cleaning effect: A few Cs-cycles clean off all carbon from the surface that has been present in graphite structure. This observation supplies further evidence for a C-Cs-compound model. Returning to the 92 eV Auger peak it is proposed due to a $NN_{Cs}L_C$ cross transition.

16. Metal Oxides

How about the metal oxides which exhibit unusually big peak shifts? Some have already been interpreted as cross transitions. The example best known is that of Al_2O_3. Fig. 10 is redrawn from the work of QUINTO and ROBERTSON /35/ who investigated the Auger spectra of Al_2O_3 and Al. They began with the oxide and gradually cleaned it off by ion bombardment. Curves a) and c) compare prominent peaks of the oxide and the metal surface respectively, b) represents an intermediate stage where both oxide and clean metal are exposed. The observed shift is as high as 15 eV. Based on an investigation of the local density of states in Al_2O_3 and Al by FOMICHEV /42/ using X-ray spectroscopy the 52 eV Auger peak was interpreted as a cross transition between aluminium and oxygen. The process is thought to be very much like that delineated by fig. 8, with the initial vacancy in an Al L-level and the emitted Auger electron being a 2p electron of oxygen. Although the facts are quite convincing the existence of cross transitions is by no means proven. BAUER still prefers the interpretation of extremely large chemical shifts of aluminium levels in Al_2O_3 /3/.

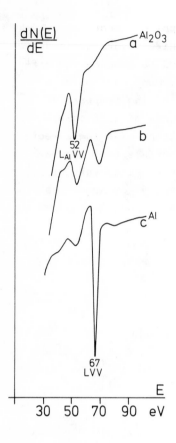

Fig.10 Comparison of LVV peaks from Al_2O_3 and Al (drawn from /35/). a) Al_2O_3, the 52 eV peak is probably due to a cross transition between the L-level of Al and oxygen band states. c) Al, b) both metal and oxide present at the surface.

The cases of the other oxides already mentioned are quite similar to describe. Among the expected features there are peaks in the wrong place. While the local density of states is generally not known it is always possible to ascribe them to cross transitions between levels that have suffered only normal chemical shifts of a few eV.

More experiments with compounds other than oxides /43/ shall be necessary to supply further evidence for the existence and the proper interpretation of composite peaks in Auger spectra. Although one should expect cross transitions in alloys, none have been found in the Cu-Ni system /44/.

17. Conclusion

It is difficult to overstate the importance which AES has already
gained in surface physics and chemistry. Often along with LEED it has
revealed most delicate information about surfaces by contributing a
qualitative chemical analysis. Profiles can be carved in surfaces
and chemically analysed by combining AES and ion bombardment techniques. First steps towards quantitative measurements have been successful and certainly will develope some kind of routine. Moreover,
there is a strong possibility that information about the electronic
structure of surfaces such as the density of states can be extracted
from AES. Future progress, however, can only comply with these pros-
pects if the different methods in surface physics are applied in a
complementary manner /28/. Some of the electron and photon impact
processes have briefly been mentioned in their relation to AES, but
other methods must be included as well. In view of the headline of
this paper future progress in Auger electron spectroscopy does have
a good chance to develope an even more sophisticated chemical surface
analysis.

Abstract

Slow Auger electrons are widely used for the identification of traces
of contaminants at solid surfaces. Although transition probabilities
for individual atoms are not known, even quantitative determinations
can successfully be conveyed by integration over all transitions initiated by the same vacancy. Auger electrons also carry information
about the electronic properties of the selvedge. Provided certain
assumptions can be made it is possible to retrieve the density of
states from the fine structure of peaks. Variation of peak shape
with chemical environment such as adsorbed species is interpreted
in terms of molecular orbitals involved in the chemical bonding. This
also leads to chemical shifts. In some cases it is feasible to assume
a finite probability for cross transitions between electronic levels
of different elements in a compound. If appropriately interpreted cross
transitions could indicate the presence of compounds.

Acknowledgement

Throughout this paper I have drawn on the experience and some of the
results of my former and present colleagues W. Gamm, H. Schiefer,
O. Albrecht, E. Lang and R. Weißmann, whose cooperation is greatly
appreciated.

123

References

1. A. BENNINGHOVEN: Appl. Physics 1, 3 (1973).

2. J.C. TRACY: in W. Dekeyser et al. (eds.): Electron Emission
    Spectroscopy. Dordrecht-Holland: D. Reidel Publishing Company,
    1973, pp 295 - 372.

3. E. BAUER: Vacuum 22, 539 (1972).

4. C.R. BRUNDLE: Surf. Sci. 27, 681 (1971).

5. R.L. GERLACH: J. Vac. Sci. Techn. 8, 599 (1971).

6. L. FIERMANS, J. VENNIK: Surf. Sci. 38, 257 (1973).

7. L.H. JENKINS, M.F. CHUNG: Surf. Sci. 26, 151 (1971).

8. M. SULEMAN, E.B. PATTINSON: J. Phys. F: Metal Physics 1, L 21
    (1971).

9. J. KIRSCHNER, P. STAIB: Phys. Letters 42 A, 335 (1973)
                       and: Appl. Physics 6, 99 (1975).

10. E.H.S. BURHOP, W.N. ASAAD: Adv. At. Mol. Phys. 8, 164 (1972).

11. J.E. HOUSTON, R.L. PARK: J. Vac. Sci. Techn. 8, 91 (1971).

12. K. SIEGBAHN, C. NORDLING, A. FAHLMAN, R. NORDBERG, K. HAMRIN,
    J. HEDMAN, G. JOHANSSON, T. BERGMARK, S.E. KARLSSON, I. LINDGREN,
    B. LINDBERG: ESCA, atomic molecular and solid state structure
    studied by means of electron spectroscopy. Uppsala: Almquist
    and Wiksells, 1967.

13. J.J. LANDER: Phys. Rev. 91, 1382 (1953).

14. E.J. McGUIRE: Phys. Rev. A 3, 1801 (1971).

15. M.F. CHUNG, L.H. JENKINS: Surf. Sci. 22, 479 (1970).

16. W.A. COGHLAN, R.E. CLANSING: A catalog of calculated Auger
    transitions for the elements, USAEC report ORNL-TM-3576, Oak
    Ridge National Laboratory, 1971.

17. L. FIERMANS, J. VENNIK: Surf. Sci. 38, 237 (1973).

18. E.N. SICKAFUS: Rev. Sci. Instr. 42, 933 (1971).

19. P. STAIB, J. KIRSCHNER: Appl. Physics 3, 421 (1974).

20. J.E. HOUSTON: Appl. Physics (1975).

21. R.L. GERLACH, A.R. Du CHARME: Surf. Sci. 32, 329 (1972).

22. K. JACOBI, J. HÖLZL: Surf. Sci. 26, 54 (1971).

23. M.P. SEAH: Surf. Sci. 32, 703 (1972).

24. W.M. MULARIE, W.T. PERIA: Surf. Sci. 26, 125 (1971).

25. H.G. MAGUIRE, P.D. AUGUSTUS: J. Phys. C $\underline{4}$, L 174 (1971).

26. G.F. AMELIO: Surf. Sci. $\underline{22}$, 301 (1970).

27. C.J. POWELL, A. MANDL: Phys. Rev. Letters $\underline{29}$, 1153 (1972).

28. J. TEJEDA, N.J. SHEVCHIK, D.W. LANGER, M. CARDONA: Phys. Rev. Letters $\underline{30}$, 370 (1973).

29. C.J. POWELL: Phys. Rev. Letters $\underline{30}$, 1179 (1973).

30. G.E. BECKER, H.D. HAGSTRUM: Surf. Sci. $\underline{30}$, 134 (1972).

31. E.N. SICKAFUS: Phys. Rev. B $\underline{7}$, 5100 (1973).

32. J.P. COAD, J.C. RIVIERE: Proc. Roy. Soc. Lond. A $\underline{331}$, 403 (1972).

33. J.T. GRANT, T.W. HAAS: Surf. Sci. $\underline{24}$, 332 (1971).

34. L. FIERMANS, J. VENNIK: Surf. Sci. $\underline{35}$, 42 (1973).

35. D.T. QUINTO, W.D. ROBERTSON: Surf. Sci. $\underline{27}$, 645 (1971).

36. D.M. ZEHNER, N. BARBULESCO, L.H. JENKINS: Surf. Sci. $\underline{34}$, 385 (1973).

37. M. SULEMAN, E.B. PATTINSON: Surf. Sci. $\underline{35}$, 75 (1973).

38. R.E. CLAUSING, D.S. EASTON: Surf. Sci. $\underline{36}$, 377 (1973).

39. F. MEYER, J.J. VRAKKING: Surf. Sci. $\underline{38}$, 275 (1973).

40. W. GAMM, K. MÜLLER: Verhandl. DPG No. 5, 463 (1972).

41. F.J. SALZANO, S. ARONSON: J. Chem. Phys. $\underline{42}$, 1323 (1965).

42. V.A. FOMICHEV: Sov. Phys. Sol. State $\underline{8}$, 2312 (1967).

43. H.H. FARRELL: Surf. Sci. $\underline{34}$, 465 (1973).

44. D.T. QUINTO, V.S. SUNDARAM, W.D. ROBERTSON: Surf. Sci. $\underline{28}$, 504 (1971).

Springer Tracts in Modern Physics

Vol. 66:
Quantum Statistics in Optics and Solid-State Physics
30 figures. III, 173 pages. 1973
Cloth DM 78,–; US $32.00   ISBN 3-540-06189-4

Vol. 67: S. FERRARA, R. GATTO, A.F. GRILLO:
**Conformal Algebra in Space-Time
and Operator Product Expansion**
II, 69 pages. 1973. Cloth DM 38,–; US $15.60
ISBN 3-540-06216-5

Vol. 68:
Solid-State Physics
77 figures. 48 tables. III, 205 pages. 1973
Cloth DM 88,–; US $36.10   ISBN 3-540-06341-2

Vol. 69:
Astrophysics
13 figures. III, 121 pages. 1973
Cloth DM 78,–; US $32.00   ISBN 3-540-06376-5

Vol. 70:
Quantum Optics
II, 135 pages. 1974. Cloth DM 77,–; US $31.60
ISBN 3-540-06630-6

Vol. 71:
Nuclear Physics
116 figures. III, 245 pages. 1974
Cloth DM 98,–; US $40.20   ISBN 3-540-06641-1

Vol. 72: D. LANGBEIN
Theory of Van der Waals Attraction
32 figures. II, 145 pages. 1974
Cloth DM 78,–; US $32.00   ISBN 3-540-06742-6

Vol. 73:
Excitons at High Density
Edited by H. Haken, S. Nikitine
With contributions by numerous experts
120 figures. IV, 303 pages. 1975
Cloth DM 97,–; US $39.80   ISBN 3-540-06943-7

Vol. 74:
Solid-State Physics
75 figures. III, 153 pages. 1974
Cloth DM 78,–; US $32.00   ISBN 3-540-06946-1

Vol. 75: R. CLAUS, L. MERTEN, J. BRANDMÜLLER:
Light Scattering by Phonon – Polaritons
55 figures. VII, 237 pages. 1975
Cloth DM 69,–; US $28.30   ISBN 3-540-07423-6

Vol. 76: H. ULLMAIER
Irreversible Properties of Type II Superconductors
Approx. 70 figures. Approx. 180 pages. 1975
Cloth DM 58,–; US $23.80   ISBN 3-540-07424-4

**Springer-Verlag
Berlin
Heidelberg
New York**

Prices are subject to change without notice

Interactions on Metal Surfaces

Edited by R. Gomer
112 figures. XI, 310 pages. 1975. (Topics in Applied
Physics, Vol. 4). Cloth DM 78,–; US $32.00
ISBN 3-540-07094-X

The purpose of this book is to acquaint the reader with
modern developments in the study of metal surfaces,
principally chemisorption. To this end, the theory of
the electronic properties of metal surfaces and the theory
of chemisorption are outlined. This part is followed by
discussions of the current experimental status of
chemisorption on well-defined single crystal planes.
Subsequent chapters describe in some detail the most
important techniques for characterizing clean and adsorb-
ate-covered surfaces, and some of the results obtained with
them. These include thermal, electron-impact and photo-
desorption, field emission and photoemission spectroscopy;
and low-energy electron diffraction and Auger analysis.
Finally, a discussion of catalysis, which transcends interactions on metal surfaces, is included it represents one of
the most important applications of chemisorption.

Prices are subject to change without notice

Applied Physics

Editorial Board: A. Benninghoven, Münster; R. Gomer,
Chicago, Ill.; F. Kneubühl, Zürich; H.K.V. Lotsch,
Heidelberg; H.J. Queisser, Stuttgart; F.P. Schäfer,
Göttingen; A. Seeger, Stuttgart; K. Shimoda, Tokyo;
T. Tamir, Brooklyn, N.Y.; H.P.J. Wijn, Eindhoven; H. Wolter,
Marburg

The purpose of this high-quality journal is to serve as a
vehicle for the publication of original papers, both
experimental and theoretical. These should contribute
new knowledge or increase understanding of phenomena,
principles, or methods relevant to applied physics. Space
is devoted to both letter and contributed papers.
Sometimes a topic may be introduced in an issue and
highlighted by a review or tutorial article (invited paper).

Sample copies as well as subscription and back-volume
information available upon request.

Please address:

Springer-Verlag, Werbeabteilung 4021
D-1000 Berlin 33, Heidelberger Platz 3
or
Springer-Verlag, New York Inc.
Promotion Department,
175 Fifth Avenue, New York, N Y 10010

Springer-Verlag
Berlin
Heidelberg
New York

Date Due

Due	Returned	Due	Returned
Sept. 22, 77			
SEP 2 5 1979			
JUN 21 1985			
MAR 14 1986			
MAY 2 1 1989			
NOV 2 0 1996	NOV 0 6 1996		
APR 0 1 1998	DEC 1 6 1997		
AUG 0 3 2001	NOV 2 7 2001		
APR 0 9 2002	OCT 0 9 2002		
NOV 1 7 2003	OCT 1 1 2002 JUN 0 4 2004		